AF541276

FUNDAMENTAL OF MECHANICAL ENGINEERING

FUNDAMENTAL OF MECHANICAL ENGINEERING

– Author –
Kulbushan Vohra
Ritika Singh

ISBN: 978-81-19283-23-1 (Hardback)

Published by : **WORDPEN ACADEMICS**
4736/23, Ansari Road, Darya Ganj,
New Delhi – 110 002
Phone: +91-011-35004336
E-mail: info@wordpenacademics.com

Printed at : **Replika Press Pvt. Ltd.**

Exclusive Distributor: ***Bio-Green Books, New Delhi***

PREFACE

Mechanical Engineering is a branch of engineering that deals with the design, analysis, manufacturing, and maintenance of mechanical systems. Mechanical engineers work on various devices and systems that involve mechanical components, including machinery, engines, vehicles, robotics, and energy systems. They apply principles of physics, mathematics, and material science to design and develop products and systems that are efficient, reliable, and safe. Mechanical engineers play a crucial role in many industries, including manufacturing, transportation, energy, construction, and healthcare. They are responsible for improving existing products and systems, as well as creating new ones that meet the needs of society. Mechanical engineers are involved in the design and development of various machines and systems that we use in our daily lives. This includes everything from small consumer products to large industrial machinery.

Mechanical engineering plays a crucial role in the manufacturing industry. Mechanical engineers design and develop tools and machines that are used in manufacturing processes to increase efficiency and reduce costs. Mechanical engineering is essential in the transportation industry, where mechanical engineers design and develop vehicles, engines, and other systems used in transportation. Mechanical engineers are involved in the development of various energy systems, including renewable energy sources like wind turbines and solar panels, as well as traditional energy sources like oil and gas. Mechanical engineering is also involved in the construction industry, where mechanical engineers design and develop various systems used in construction projects, including heating, ventilation, and air conditioning systems. Mechanical engineering is an innovative and research-oriented field that is constantly evolving. Mechanical engineers are continually developing new technologies and processes to improve existing systems and create new ones. Mechanical engineering plays a crucial role in various industries, and its contributions are essential to our modern way of life. It is a dynamic and exciting field that offers many opportunities for innovation, growth, and development.

The book starts with an introduction to mechanical engineering, its importance, and the various fields of study within the discipline. The book covers the fundamental principles of mechanics, including force, motion, energy, and momentum and describes the history of Aristotelian physics and its impact on modern mechanical engineering. It explains the basic principles of thermodynamics, including the laws of thermodynamics, heat transfer, and thermodynamic cycles and also explains the different methods of power transmission, including gears, pulleys, belts, and chains. It describes the basics of boilers, including their types, components, and applications and also covers various types of engines, including internal combustion engines and steam engines. The book explains the basic principles of hydraulics, including the properties of fluids, fluid mechanics, and hydraulic systems. It also covers the design and operation of water turbines. It emphasizes the design process for mechanical systems, including concept development, analysis, and prototyping and also covers the covers the principles of material handling, including the types of equipment used, safety considerations, and the design of material handling systems. It is an excellent resource for students and professionals alike who are interested in this exciting and essential field.

We are grateful to all those persons as well as various books, manuals, periodicals, magazines, journals etc. that helped in the preparation of this book. In spite of the best efforts, it is possible that some errors may have occurred into the compilation and editing of the book. Further queries, constructive suggestions and criticisms for the improvement of the book are always welcomed and shall be thankfully acknowledged.

Kulbushan Vohra

Ritika Singh

CONTENTS

1

INTRODUCTION

Technically, mechanical engineering is the application of the principles and problem-solving techniques of engineering from design to manufacturing to the marketplace for any object. Mechanical engineers analyze their work using the principles of motion, energy, and force—ensuring that designs function safely, efficiently, and reliably, all at a competitive cost.

Mechanical engineers make a difference. That's because mechanical engineering careers center on creating technologies to meet human needs. Virtually every product or service in modern life has probably been touched in some way by a mechanical engineer to help humankind.

This includes solving today's problems and creating future solutions in health care, energy, transportation, world hunger, space exploration, climate change, and more.

Being ingrained in many challenges and innovations across many fields means a mechanical engineering education is versatile. To meet this broad demand, mechanical engineers may design a component, a machine, a system, or a process. This ranges from the macro to the micro, from the largest systems like cars and satellites to the smallest components like sensors and switches. Anything that needs to be manufactured—indeed, anything with moving parts—needs the expertise of a mechanical engineer. Become a mechanical engineer.

Mechanical engineering is an engineering branch that combines engineering physics and mathematics principles with materials science, to design, analyze, manufacture, and maintain mechanical systems. It is one of the oldest and broadest of the engineering branches. The mechanical engineering field requires an understanding of core areas including mechanics, dynamics, thermodynamics, materials science, structural analysis, and electricity. In addition to these core principles, mechanical engineers use tools such as computer-aided design (CAD), computer-aided manufacturing (CAM), and product lifecycle management to design and analyze manufacturing plants, industrial equipment and machinery, heating and cooling systems, transport systems, aircraft, watercraft, robotics, medical devices, weapons, and others. It is the branch of engineering that involves the design, production, and operation of machinery.

Mechanical engineering emerged as a field during the Industrial Revolution in Europe in the 18th century; however, its development can be traced back several thousand years around the world. In the 19th century, developments in physics led to the development of mechanical engineering science. The field has continually evolved to incorporate advancements; today mechanical engineers are pursuing developments in such areas as composites, mechatronics, and nanotechnology. It also overlaps with aerospace engineering, metallurgical engineering, civil engineering, electrical engineering, manufacturing engineering, chemical engineering, industrial engineering, and other engineering disciplines to varying amounts. Mechanical engineers may also work in the field of biomedical engineering, specifically with biomechanics, transport phenomena, biomechatronics, bionanotechnology, and modelling of biological systems.

HISTORY

Mechanical engineering is a discipline centered around the concept of using force multipliers, moving components, and machines. It utilizes knowledge of mathematics, physics, materials sciences, and engineering technologies. It is one of the oldest and broadest of the engineering disciplines.

Engineering arose in early civilization as a general discipline for the creation of large scale structures such as irrigation, architecture, and military projects. Advances in food production through irrigation allowed a portion of the population to become specialists in Ancient Babylon.

All six of the classic simple machines were known in the ancient Near East. The wedge and the inclined plane (ramp) were known since prehistoric times. The wheel, along with the wheel and axle mechanism, was invented in Mesopotamia (modern Iraq) during the 5th millennium BC. The lever mechanism first appeared around 5,000 years ago in the Near East, where it was used in a simple balance scale, and to move large objects in ancient Egyptian technology. The lever was also used in the shadoof water-lifting device, the first crane machine, which appeared in Mesopotamia circa 3000 BC, and then in ancient Egyptian technology circa

2000 BC. The earliest evidence of pulleys date back to Mesopotamia in the early 2nd millennium BC, and ancient Egypt during the Twelfth Dynasty (1991-1802 BC). The screw, the last of the simple machines to be invented, first appeared in Mesopotamia during the Neo-Assyrian period (911-609) BC. The Egyptian pyramids were built using three of the six simple machines, the inclined plane, the wedge, and the lever, to create structures like the Great Pyramid of Giza.

The Assyrians were notable in their use of metallurgy and incorporation of iron weapons. Many of their advancements were in military equipment. They were not the first to develop them, but did make advancements on the wheel and the chariot. They made use of pivot-able axles on their wagons, allowing easy turning. They were also one of the first armies to use the move-able siege tower and battering ram.

The application of mechanical engineering can be seen in the archives of various ancient societies. The pulley appeared in Mesopotamia in 1,500 BC, improving water transportation. German Archaeologist Robert Koldewey found that the Hanging Gardens likely used a mechanical pump powered by these pulleys to transport water to the roof gardens. The Mesopotamians would advance even further by replacing "the substitution of continuous for intermittent motion, and the rotary for back-and-forth motion" by 1,200 BC.

The Sakia was developed in the kingdom of Kush during the 4th century BC. It lifted water 3 to 8 metres with less expenditure of labor and time. Reservoirs in the form of Hafirs were developed in Kush to store water and boost irrigation. Bloomeries and blast furnaces were developed during the seventh century BC in Meroe. Kushite sundials applied mathematics in the form of advanced trigonometry.

In Ancient Egypt, the screw pump is another example of the use of engineering to boost efficiency of water transportation. Although the Early Egyptians built colossal structures such as the pyramids, they did not develop pulleys for the lifting of heavy stone, and used the wheel very little.

The earliest practical water-powered machines, the water wheel and watermill, first appeared in the Persian Empire, in what are now Iraq and Iran, by the early 4th century BC.

In Ancient Greece, Archimedes (287–212 BC) developed several key theories in the field of mechanical engineering including mechanical advantage, the Law of the Lever, and his name sake, Archimedes' law. In Ptolematic Egypt, the Museum of Alexandria developed crane pulleys with block and tackles to lift stones. These cranes were powered with human tread wheels and were based on earlier Mesopotamian water-pulley systems. The Greeks would later develop mechanical artillery independently of the Chinese. The first of these would fire darts, but advancements allowed for stone to be tossed at enemy fortifications or formations.

The geared Antikythera mechanism is an example of ancient mechanical engineering.

Late Antiquity to early Middle Ages

In Roman Egypt, Heron of Alexandria (c. 10–70 AD) created the first steam-powered device, the Aeolipile. The first of its kind, it did not have the capability to move or power anything but its own rotation.

In China, Zhang Heng (78–139 AD) improved a water clock and invented a seismometer. Ma Jun (200–265 AD) invented a chariot with differential gears.

Leo the Philosopher is noted to have worked on a signal system using clocks in the Byzantine Empire in 850, connecting Constantinople with the Cicilian Frontier and was a continuation of the complex city clocks in Eastern Rome. These grand machines diffused into the Arabian Empire under Harun al-Rashid.

Another grand mechanical device was the Organ, which was reintroduced in 757 when Constantine V gifted one to Pepin the short.

With the exception of a few machines, engineering and science stagnated in the West due to the collapse of the Roman Empire during late antiquity.

Middle Ages

During the Islamic Golden Age (7th to 15th century), Muslim inventors made remarkable contributions in the field of mechanical technology. Al-Jazari, who was one of them, wrote his famous Book of Knowledge of Ingenious Mechanical Devices in 1206 and presented many mechanical designs.

The earliest practical wind-powered machines, the windmill and wind pump, first appeared in the Muslim world during the Islamic Golden Age, in what are now Iran, Afghanistan, and Pakistan, by the 9th century AD. The earliest practical steam-powered machine was a steam jack driven by a steam turbine, described in 1551 by Taqi al-Din Muhammad ibn Ma'ruf in Ottoman Egypt.

The cotton gin was invented in India by the 6th century AD, and the spinning wheel was invented in the Islamic world by the early 11th century, both of which were fundamental to the growth of the cotton industry. The spinning wheel was also a precursor to the spinning jenny, which was a key development during the early Industrial Revolution in the 18th century.

The earliest programmable machines were developed in the Muslim world. A music sequencer, a programmable musical instrument, was the earliest type of programmable machine. The first music sequencer was an automated flute player invented by the Banu Musa brothers, described in their Book of Ingenious Devices, in the 9th century. In 1206, Al-Jazari invented programmable automata/ robots. He described four automaton musicians, including drummers operated by a programmable drum machine, where they could be made to play different rhythms and different drum patterns. The castle clock, a hydropowered mechanical astronomical clock invented by Al-Jazari, was the first programmable analog computer.

The medieval Chinese horologist and engineer Su Song (1020–1101 AD) incorporated an escapement mechanism into his astronomical clock tower two centuries before escapement devices were found in medieval European clocks and also invented the world's first known endless power-transmitting chain drive.

The Middle Ages saw the wide spread adoption of machines to aid in labor. The many rivers of England and northern Europe allowed the power of moving water to be utilized. The water-mill became instrumental in the production of many goods such as food, fabric, leathers, and papers. These machines used were some of the first to use cogs and gears, which greatly increased the mills productivity. The camshaft allowed rotational force to be converted into directional force. Less significantly, tides of bodies of water were also harnessed.

Wind-power later became the new source of energy in Europe, supplementing the water mill. This advancement moved out of Europe into the Middle East during the Crusades.

Metallurgy advanced by a large degree during the Middle Ages, with higher quality iron allowing for more sturdy constructions and designs. Mills and mechanical power provided a consistent supply of trip-hammer strikes and air from the bellows.

Renaissance and Scientific Revolution

Leonardo da Vinci was a notable engineer, designing and studying many mechanical systems that were focused around transportation and warfare His designs would later be compared to early aircraft design.

Although wind power provided a source of energy away from riverside estate and saw massive improvements in its harnessing, it could not replace the consistent and strong power of the watermill. Water would remain the primary source of power of pre-industrial urban industry through the Renaissance.

In the 17th century, during the Scientific Revolution, important breakthroughs in the foundations of mechanical engineering occurred in England and the

Continent. The Dutch mathematician and physicist Christiaan Huygens invented the pendulum clock in 1657, which was the first reliable timekeeper for almost 300 years, and published a work dedicated to clock designs and the theory behind them. Isaac Newton formulated Newton's Laws of Motion and developed the calculus, which would become the mathematical basis of physics. Newton was reluctant to publish his works for years, but he was finally persuaded to do so by his colleagues, such as Edmond Halley. Gottfried Wilhelm Leibniz is also credited with developing the calculus independently of Newton during this time period.

Industrial Revolution

At the end of the Renaissance, scientists and engineers were beginning to experiment with steam power. Most of the early apparatuses faced problems of low horsepower, inefficiency, or danger. The need arose for an effective and economical power source because of the flooding of deep-mines in England, which could not be pumped out using alternative methods. The first working design was Thomas Savery's 1698 patent. He continuously worked on improving and marketing the invention across England. At the same time, others were working on improvements to Savery's design, which did not transfer heat effectively.

Thomas Newcomen would take all the advancements of the engineers and develop the Newcomen Atmospheric Engine. This new design would greatly reduce heat loss, move water directly from the engine, and allow variety of proportions to be built in.

The Industrial Revolution brought steam powered factories utilizing mechanical engineering concepts. These advances allowed an incredible increase in production scale, numbers, and efficiency.

During the 19th century, material sciences advances had begun to allow implementation of steam engines into Steam Locomotives and Steam-Powered Ships, quickly increasing the speed at which people and goods could move across the world. The reason for these advances were the machine tools were developed in England, Germany, and Scotland. These allowed mechanical engineering to develop as a separate field within engineering. They brought with them manufacturing machines and the engines to power them.

At the near end of the Industrial Revolution, internal combustion engine technology brought with it the piston airplane and automobile. Aerospace Engineering would develop in the early 20th century as an offshoot of mechanical engineering, eventually incorporating rocketry.

Modern Age

With the advents of computers in the 20th century, more precise design and manufacturing methods were available to engineers. The rise of CAD software has reduced design times and allowed for precision manufacturing. Engineers are able to simulate the forces and stresses of designs through computer programs. Automated and Computerized manufacturing allowed many new fields to emerge from Mechanical Engineering such as Industrial Engineering. Although a majority of automobiles remain to be gas powered, electric vehicles have risen as a feasible alternative.

Because of the increased complexity of engineering projects, many disciplines of engineer collaborate and specialize in sub fields. One of these collaborations is the field of robotics, in which electrical engineers, computer engineers, and mechanical engineers can specialize in and work together. Mechanical Engineering is the most popular of all the engineering fields for college majors in the 21st century.

Professional associations

The first British professional society of mechanical engineers was formed in 1847 Institution of Mechanical Engineers, thirty years after the civil engineers formed the first such professional society Institution of Civil Engineers.

In the United States, the American Society of Mechanical Engineers (ASME) was formed in 1880, becoming the third such professional engineering society, after the American Society of Civil Engineers (1852) and the American Institute of Mining Engineers (1871).

Education

The first schools in the United States to offer an mechanical engineering education were the United States Military Academy in 1817, an institution now known as Norwich University in 1819, and Rensselaer Polytechnic Institute in 1825. Education in mechanical engineering has historically been based on a strong foundation in mathematics and science.

In the 20th century, many governments began regulating both the title of engineer and the practice of engineering, requiring a degree from an accredited university and to pass a qualifying test.

Education

Degrees in mechanical engineering are offered at various universities worldwide. Mechanical engineering programs typically take four to five years of study depending on the place and university and result in a Bachelor of Engineering (B.Eng. or B.E.), Bachelor of Science (B.Sc. or B.S.), Bachelor of Science Engineering (B.Sc.Eng.), Bachelor of Technology (B.Tech.), Bachelor of Mechanical Engineering (B.M.E.), or Bachelor of Applied Science (B.A.Sc.) degree, in or with emphasis in mechanical engineering. In Spain, Portugal and most of South America, where neither B.S. nor B.Tech. programs have been adopted, the formal name for the degree is "Mechanical Engineer", and the course work is based on five or six years of training. In Italy the course work is based on five years of education, and training, but in order to qualify as an Engineer one has to pass a state exam at the end of the course. In Greece, the coursework is based on a five-year curriculum and the requirement of a 'Diploma' Thesis, which upon completion a 'Diploma' is awarded rather than a B.Sc.

In the United States, most undergraduate mechanical engineering programs are accredited by the Accreditation Board for Engineering and Technology (ABET) to ensure similar course requirements and standards among universities. The ABET web site lists 302 accredited mechanical engineering programs as of 11 March 2014. Mechanical engineering programs in Canada are accredited by the Canadian Engineering Accreditation Board (CEAB), and most other countries offering engineering degrees have similar accreditation societies.

In Australia, mechanical engineering degrees are awarded as Bachelor of Engineering (Mechanical) or similar nomenclature, although there are an increasing number of specialisations. The degree takes four years of full-time study to achieve. To ensure quality in engineering degrees, Engineers Australia accredits engineering degrees awarded by Australian universities in accordance with the global Washington Accord. Before the degree can be awarded, the student must complete at least 3 months of on the job work experience in an engineering firm. Similar systems are also present in South Africa and are overseen by the Engineering Council of South Africa (ECSA).

In India, to become an engineer, one needs to have an engineering degree like a B.Tech or B.E, have a diploma in engineering, or by completing a course in an engineering trade like fitter from the Industrial Training Institute (ITIs) to receive a "ITI Trade Certificate" and also pass the All India Trade Test (AITT) with an engineering trade conducted by the National Council of Vocational Training (NCVT) by which one is awarded a "National Trade Certificate". A similar system is used in Nepal.

Some mechanical engineers go on to pursue a postgraduate degree such as a Master of Engineering, Master of Technology, Master of Science, Master of

Engineering Management (M.Eng.Mgt. or M.E.M.), a Doctor of Philosophy in engineering (Eng.D. or Ph.D.) or an engineer's degree. The master's and engineer's degrees may or may not include research. The Doctor of Philosophy includes a significant research component and is often viewed as the entry point to academia. The Engineer's degree exists at a few institutions at an intermediate level between the master's degree and the doctorate.

Coursework

Standards set by each country's accreditation society are intended to provide uniformity in fundamental subject material, promote competence among graduating engineers, and to maintain confidence in the engineering profession as a whole. Engineering programs in the U.S., for example, are required by ABET to show that their students can "work professionally in both thermal and mechanical systems areas." The specific courses required to graduate, however, may differ from program to program. Universities and institutes of technology will often combine multiple subjects into a single class or split a subject into multiple classes, depending on the faculty available and the university's major area(s) of research.

The fundamental subjects required for mechanical engineering usually include:

- Mathematics (in particular, calculus, differential equations, and linear algebra)
- Basic physical sciences (including physics and chemistry)
- Statics and dynamics
- Strength of materials and solid mechanics
- Materials engineering, composites
- Thermodynamics, heat transfer, energy conversion, and HVAC
- Fuels, combustion, internal combustion engine
- Fluid mechanics (including fluid statics and fluid dynamics)
- Mechanism and Machine design (including kinematics and dynamics)
- Instrumentation and measurement
- Manufacturing engineering, technology, or processes
- Vibration, control theory and control engineering
- Hydraulics and Pneumatics
- Mechatronics and robotics
- Engineering design and product design

- Drafting, computer-aided design (CAD) and computer-aided manufacturing (CAM)

Mechanical engineers are also expected to understand and be able to apply basic concepts from chemistry, physics, tribology, chemical engineering, civil engineering, and electrical engineering. All mechanical engineering programs include multiple semesters of mathematical classes including calculus, and advanced mathematical concepts including differential equations, partial differential equations, linear algebra, abstract algebra, and differential geometry, among others.

In addition to the core mechanical engineering curriculum, many mechanical engineering programs offer more specialized programs and classes, such as control systems, robotics, transport and logistics, cryogenics, fuel technology, automotive engineering, biomechanics, vibration, optics and others, if a separate department does not exist for these subjects.

Most mechanical engineering programs also require varying amounts of research or community projects to gain practical problem-solving experience. In the United States it is common for mechanical engineering students to complete one or more internships while studying, though this is not typically mandated by the university. Cooperative education is another option. Future work skills research puts demand on study components that feed student's creativity and innovation.

Job duties

Mechanical engineers research, design, develop, build, and test mechanical and thermal devices, including tools, engines, and machines.

Mechanical engineers typically do the following:

- Analyze problems to see how mechanical and thermal devices might help solve the problem.
- Design or redesign mechanical and thermal devices using analysis and computer-aided design.
- Develop and test prototypes of devices they design.
- Analyze the test results and change the design as needed.
- Oversee the manufacturing process for the device.
- Manage a team of professionals in specialized fields like mechanical drafting and designing, prototyping, 3D printing or/and CNC Machines specialists.
- Mechanical engineers design and oversee the manufacturing of many products ranging from medical devices to new batteries. They also

design power-producing machines such as electric generators, internal combustion engines, and steam and gas turbines as well as power-using machines, such as refrigeration and air-conditioning systems.

- Like other engineers, mechanical engineers use computers to help create and analyze designs, run simulations and test how a machine is likely to work.

License and regulation

Engineers may seek license by a state, provincial, or national government. The purpose of this process is to ensure that engineers possess the necessary technical knowledge, real-world experience, and knowledge of the local legal system to practice engineering at a professional level. Once certified, the engineer is given the title of Professional Engineer (United States, Canada, Japan, South Korea, Bangladesh and South Africa), Chartered Engineer (in the United Kingdom, Ireland, India and Zimbabwe), Chartered Professional Engineer (in Australia and New Zealand) or European Engineer (much of the European Union).

In the U.S., to become a licensed Professional Engineer (PE), an engineer must pass the comprehensive FE (Fundamentals of Engineering) exam, work a minimum of 4 years as an Engineering Intern (EI) or Engineer-in-Training (EIT), and pass the "Principles and Practice" or PE (Practicing Engineer or Professional Engineer) exams. The requirements and steps of this process are set forth by the National Council of Examiners for Engineering and Surveying (NCEES), composed of engineering and land surveying licensing boards representing all U.S. states and territories.

In the UK, current graduates require a BEng plus an appropriate master's degree or an integrated MEng degree, a minimum of 4 years post graduate on the job competency development and a peer reviewed project report to become a Chartered Mechanical Engineer (CEng, MIMechE) through the Institution of Mechanical Engineers. CEng MIMechE can also be obtained via an examination route administered by the City and Guilds of London Institute.

In most developed countries, certain engineering tasks, such as the design of bridges, electric power plants, and chemical plants, must be approved by a professional engineer or a chartered engineer. "Only a licensed engineer, for instance, may prepare, sign, seal and submit engineering plans and drawings to a public authority for approval, or to seal engineering work for public and private clients." This requirement can be written into state and provincial legislation, such as in the Canadian provinces, for example the Ontario or Quebec's Engineer Act.

In other countries, such as Australia, and the UK, no such legislation exists; however, practically all certifying bodies maintain a code of ethics independent of legislation, that they expect all members to abide by or risk expulsion.

Further information: FE Exam, Professional Engineer, Incorporated Engineer, Washington Accord, and Regulation and licensure in engineering

Salaries and workforce statistics

The total number of engineers employed in the U.S. in 2015 was roughly 1.6 million. Of these, 278,340 were mechanical engineers (17.28%), the largest discipline by size. In 2012, the median annual income of mechanical engineers in the U.S. workforce was $80,580. The median income was highest when working for the government ($92,030), and lowest in education ($57,090). In 2014, the total number of mechanical engineering jobs was projected to grow 5% over the next decade. As of 2009, the average starting salary was $58,800 with a bachelor's degree.

Subdisciplines

The field of mechanical engineering can be thought of as a collection of many mechanical engineering science disciplines. Several of these subdisciplines which are typically taught at the undergraduate level are listed below, with a brief explanation and the most common application of each. Some of these subdisciplines are unique to mechanical engineering, while others are a combination of mechanical engineering and one or more other disciplines. Most work that a mechanical engineer does uses skills and techniques from several of these subdisciplines, as well as specialized subdisciplines. Specialized subdisciplines, as used in this article, are more likely to be the subject of graduate studies or on-the-job training than undergraduate research. Several specialized subdisciplines are discussed in this section.

Mechanics

Mechanics is, in the most general sense, the study of forces and their effect upon matter. Typically, engineering mechanics is used to analyze and predict the acceleration and deformation (both elastic and plastic) of objects under known forces (also called loads) or stresses. Subdisciplines of mechanics include

Statics, the study of non-moving bodies under known loads, how forces affect static bodies

Dynamics the study of how forces affect moving bodies. Dynamics includes kinematics (about movement, velocity, and acceleration) and kinetics (about forces and resulting accelerations).

Mechanics of materials, the study of how different materials deform under various types of stress

Fluid mechanics, the study of how fluids react to forces

Kinematics, the study of the motion of bodies (objects) and systems (groups of objects), while ignoring the forces that cause the motion. Kinematics is often used in the design and analysis of mechanisms.

Continuum mechanics, a method of applying mechanics that assumes that objects are continuous (rather than discrete)

Mechanical engineers typically use mechanics in the design or analysis phases of engineering. If the engineering project were the design of a vehicle, statics might be employed to design the frame of the vehicle, in order to evaluate where the stresses will be most intense. Dynamics might be used when designing the car's engine, to evaluate the forces in the pistons and cams as the engine cycles. Mechanics of materials might be used to choose appropriate materials for the frame and engine. Fluid mechanics might be used to design a ventilation system for the vehicle (see HVAC), or to design the intake system for the engine.

MECHATRONICS AND ROBOTICS

Training FMS with learning robot SCORBOT-ER 4u, workbench CNC Mill and CNC Lathe

Mechatronics is a combination of mechanics and electronics. It is an interdisciplinary branch of mechanical engineering, electrical engineering and software engineering that is concerned with integrating electrical and mechanical engineering to create hybrid automation systems. In this way, machines can be automated through the use of electric motors, servo-mechanisms, and other electrical systems in conjunction with special software. A common example of a mechatronics system is a CD-ROM drive. Mechanical systems open and close the drive, spin the CD and move the laser, while an optical system reads the data on the CD and converts it to bits. Integrated software controls the process and communicates the contents of the CD to the computer.

Robotics is the application of mechatronics to create robots, which are often used in industry to perform tasks that are dangerous, unpleasant, or repetitive. These robots may be of any shape and size, but all are preprogrammed and interact physically with the world. To create a robot, an engineer typically employs kinematics (to determine the robot's range of motion) and mechanics (to determine the stresses within the robot).

Robots are used extensively in industrial automation engineering. They allow businesses to save money on labor, perform tasks that are either too dangerous or too precise for humans to perform them economically, and to ensure better quality. Many companies employ assembly lines of robots, especially in Automotive Industries and some factories are so robotized that they can run by themselves. Outside the factory, robots have been employed in bomb disposal, space exploration, and many other fields. Robots are also sold for various residential applications, from recreation to domestic applications.

Structural analysis

Structural analysis is the branch of mechanical engineering (and also civil engineering) devoted to examining why and how objects fail and to fix the objects and their performance. Structural failures occur in two general modes: static failure, and fatigue failure. Static structural failure occurs when, upon being loaded (having a force applied) the object being analyzed either breaks or is deformed plastically, depending on the criterion for failure. Fatigue failure occurs when an object fails after a number of repeated loading and unloading cycles. Fatigue failure occurs because of imperfections in the object: a microscopic crack on the surface of the object, for instance, will grow slightly with each cycle (propagation) until the crack is large enough to cause ultimate failure.

Failure is not simply defined as when a part breaks, however; it is defined as when a part does not operate as intended. Some systems, such as the perforated top sections of some plastic bags, are designed to break. If these systems do not break, failure analysis might be employed to determine the cause.

Structural analysis is often used by mechanical engineers after a failure has occurred, or when designing to prevent failure. Engineers often use online documents and books such as those published by ASM to aid them in determining the type of failure and possible causes.

Once theory is applied to a mechanical design, physical testing is often performed to verify calculated results. Structural analysis may be used in an office when designing parts, in the field to analyze failed parts, or in laboratories where parts might undergo controlled failure tests.

Thermodynamics and thermo-science

Thermodynamics is an applied science used in several branches of engineering, including mechanical and chemical engineering. At its simplest, thermodynamics is the study of energy, its use and transformation through a system. Typically, engineering thermodynamics is concerned with changing energy from one form to another. As an example, automotive engines convert chemical energy (enthalpy) from the fuel into heat, and then into mechanical work that eventually turns the wheels.

Thermodynamics principles are used by mechanical engineers in the fields of heat transfer, thermofluids, and energy conversion. Mechanical engineers use thermo-science to design engines and power plants, heating, ventilation, and air-conditioning (HVAC) systems, heat exchangers, heat sinks, radiators, refrigeration, insulation, and others.

Design and drafting

Drafting or technical drawing is the means by which mechanical engineers design products and create instructions for manufacturing parts. A technical drawing can be a computer model or hand-drawn schematic showing all the dimensions necessary to manufacture a part, as well as assembly notes, a list of required materials, and other pertinent information. A U.S. mechanical engineer or skilled worker who creates technical drawings may be referred to as a drafter or draftsman. Drafting has historically been a two-dimensional process, but computer-aided design (CAD) programs now allow the designer to create in three dimensions.

Instructions for manufacturing a part must be fed to the necessary machinery, either manually, through programmed instructions, or through the use of a computer-aided manufacturing (CAM) or combined CAD/CAM program. Optionally, an engineer may also manually manufacture a part using the technical drawings. However, with the advent of computer numerically controlled (CNC) manufacturing, parts can now be fabricated without the need for constant technician input. Manually manufactured parts generally consist of spray coatings, surface finishes, and other processes that cannot economically or practically be done by a machine.

Drafting is used in nearly every subdiscipline of mechanical engineering, and by many other branches of engineering and architecture. Three-dimensional models created using CAD software are also commonly used in finite element analysis (FEA) and computational fluid dynamics (CFD).

Modern tools

Many mechanical engineering companies, especially those in industrialized nations, have begun to incorporate computer-aided engineering (CAE) programs into their existing design and analysis processes, including 2D and 3D solid modeling computer-aided design (CAD). This method has many benefits, including easier and more exhaustive visualization of products, the ability to create virtual assemblies of parts, and the ease of use in designing mating interfaces and tolerances.

Other CAE programs commonly used by mechanical engineers include product lifecycle management (PLM) tools and analysis tools used to perform complex simulations. Analysis tools may be used to predict product response to expected loads, including fatigue life and manufacturability. These tools include finite element analysis (FEA), computational fluid dynamics (CFD), and computer-aided manufacturing (CAM).

Using CAE programs, a mechanical design team can quickly and cheaply iterate the design process to develop a product that better meets cost, performance, and other constraints. No physical prototype need be created until the design nears completion, allowing hundreds or thousands of designs to be evaluated, instead of a relative few. In addition, CAE analysis programs can model complicated physical phenomena which cannot be solved by hand, such as viscoelasticity, complex contact between mating parts, or non-Newtonian flows.

As mechanical engineering begins to merge with other disciplines, as seen in mechatronics, multidisciplinary design optimization (MDO) is being used with other CAE programs to automate and improve the iterative design process. MDO tools wrap around existing CAE processes, allowing product evaluation to continue even after the analyst goes home for the day. They also utilize sophisticated optimization algorithms to more intelligently explore possible designs, often finding better, innovative solutions to difficult multidisciplinary design problems.

AREAS OF RESEARCH

Mechanical engineers are constantly pushing the boundaries of what is physically possible in order to produce safer, cheaper, and more efficient machines and mechanical systems. Some technologies at the cutting edge of mechanical engineering are listed below (see also exploratory engineering).

Micro electro-mechanical systems (MEMS)

Micron-scale mechanical components such as springs, gears, fluidic and heat transfer devices are fabricated from a variety of substrate materials such as silicon, glass and polymers like SU8. Examples of MEMS components are the accelerometers that are used as car airbag sensors, modern cell phones, gyroscopes for precise positioning and microfluidic devices used in biomedical applications.

Friction stir welding (FSW)

Friction stir welding, a new type of welding, was discovered in 1991 by The Welding Institute (TWI). The innovative steady state (non-fusion) welding technique joins materials previously un-weldable, including several aluminum alloys. It plays an important role in the future construction of airplanes, potentially replacing rivets. Current uses of this technology to date include welding the seams of the aluminum main Space Shuttle external tank, Orion Crew Vehicle, Boeing Delta II and Delta IV Expendable Launch Vehicles and the SpaceX Falcon 1 rocket, armor plating for amphibious assault ships, and welding the wings and fuselage panels of the new Eclipse 500 aircraft from Eclipse Aviation among an increasingly growing pool of uses.

Composites

Composites or composite materials are a combination of materials which provide different physical characteristics than either material separately. Composite material research within mechanical engineering typically focuses on designing (and, subsequently, finding applications for) stronger or more rigid materials while attempting to reduce weight, susceptibility to corrosion, and other undesirable factors. Carbon fiber reinforced composites, for instance, have been used in such diverse applications as spacecraft and fishing rods.

Mechatronics

Mechatronics is the synergistic combination of mechanical engineering, electronic engineering, and software engineering. The discipline of mechatronics began as a way to combine mechanical principles with electrical engineering. Mechatronic concepts are used in the majority of electro-mechanical systems. Typical electro-mechanical sensors used in mechatronics are strain gauges, thermocouples, and pressure transducers.

Nanotechnology

At the smallest scales, mechanical engineering becomes nanotechnology—one speculative goal of which is to create a molecular assembler to build molecules and materials via mechanosynthesis. For now that goal remains within exploratory engineering. Areas of current mechanical engineering research in nanotechnology include nanofilters, nanofilms, and nanostructures, among others.

Finite element analysis

Finite Element Analysis is a computational tool used to estimate stress, strain, and deflection of solid bodies. It uses a mesh setup with user-defined sizes to measure physical quantities at a node. The more nodes there are, the higher the precision. This field is not new, as the basis of Finite Element Analysis (FEA) or Finite Element Method (FEM) dates back to 1941. But the evolution of computers has made FEA/FEM a viable option for analysis of structural problems. Many commercial codes such as NASTRAN, ANSYS, and ABAQUS are widely used in industry for research and the design of components. Some 3D modeling and CAD software packages have added FEA modules. In the recent times, cloud simulation platforms like SimScale are becoming more common.

Other techniques such as finite difference method (FDM) and finite-volume method (FVM) are employed to solve problems relating heat and mass transfer, fluid flows, fluid surface interaction, etc.

BIOMECHANICS

Biomechanics is the application of mechanical principles to biological systems, such as humans, animals, plants, organs, and cells. Biomechanics also aids in creating prosthetic limbs and artificial organs for humans. Biomechanics is closely related to engineering, because it often uses traditional engineering sciences to analyze biological systems. Some simple applications of Newtonian mechanics and/or materials sciences can supply correct approximations to the mechanics of many biological systems.

In the past decade, reverse engineering of materials found in nature such as bone matter has gained funding in academia. The structure of bone matter is optimized for its purpose of bearing a large amount of compressive stress per unit weight. The goal is to replace crude steel with bio-material for structural design.

Over the past decade the Finite element method (FEM) has also entered the Biomedical sector highlighting further engineering aspects of Biomechanics. FEM has since then established itself as an alternative to in vivo surgical assessment and gained the wide acceptance of academia. The main advantage of Computational Biomechanics lies in its ability to determine the endo-anatomical response of an anatomy, without being subject to ethical restrictions. This has led FE modelling to the point of becoming ubiquitous in several fields of Biomechanics while several projects have even adopted an open source philosophy (e.g. BioSpine).

Computational fluid dynamics

Computational fluid dynamics, usually abbreviated as CFD, is a branch of fluid mechanics that uses numerical methods and algorithms to solve and analyze problems that involve fluid flows. Computers are used to perform the calculations required to simulate the interaction of liquids and gases with surfaces defined by boundary conditions. With high-speed supercomputers, better solutions can be achieved. Ongoing research yields software that improves the accuracy and speed of complex simulation scenarios such as turbulent flows. Initial validation of such software is performed using a wind tunnel with the final validation coming in full-scale testing, e.g. flight tests.

Acoustical engineering

Acoustical engineering is one of many other sub-disciplines of mechanical engineering and is the application of acoustics. Acoustical engineering is the study of Sound and Vibration. These engineers work effectively to reduce noise pollution in mechanical devices and in buildings by soundproofing or removing sources of unwanted noise. The study of acoustics can range from designing a more efficient hearing aid, microphone, headphone, or recording studio to enhancing the sound quality of an orchestra hall. Acoustical engineering also deals with the vibration of different mechanical systems.

What do mechanical engineers do?

Mechanical engineering combines creativity, knowledge and analytical tools to complete the difficult task of shaping an idea into reality.

This transformation happens at the personal scale, affecting human lives on a level we can reach out and touch like robotic prostheses. It happens on the local scale, affecting people in community-level spaces, like with agile interconnected microgrids. And it happens on bigger scales, like with advanced power systems, through engineering that operates nationwide or across the globe.

Mechanical engineers have an enormous range of opportunity and their education mirrors this breadth of subjects. Students concentrate on one area while strengthening analytical and problem-solving skills applicable to any engineering situation.

Disciplines within mechanical engineering include but are not limited to:

- Acoustics
- Aerospace
- Automation
- Automotive
- Autonomous Systems
- Biotechnology

- Composites
- Computer Aided Design (CAD)
- Control Systems
- Cyber security
- Design
- Energy
- Ergonomics
- Human health
- Manufacturing and additive manufacturing
- Mechanics
- Nanotechnology
- Production planning
- Robotics
- Structural analysis

Technology itself has also shaped how mechanical engineers work and the suite of tools has grown quite powerful in recent decades. Computer-aided engineering (CAE) is an umbrella term that covers everything from typical CAD techniques to computer-aided manufacturing to computer-aided engineering, involving finite element analysis (FEA) and computational fluid dynamics (CFD). These tools and others have further broadened the horizons of mechanical engineering.

WHAT CAREERS ARE THERE IN MECHANICAL ENGINEERING?

Society depends on mechanical engineering. The need for this expertise is great in so many fields, and as such, there is no real limit for the freshly minted mechanical engineer. Jobs are always in demand, particularly in the automotive, aerospace, electronics, biotechnology, and energy industries.

Here are a handful of mechanical engineering fields.

In statics, research focuses on how forces are transmitted to and throughout a structure. Once a system is in motion, mechanical engineers look at dynamics, or what velocities, accelerations and resulting forces come into play. Kinematics then examines how a mechanism behaves as it moves through its range of motion.

Materials science delves into determining the best materials for different applications. A part of that is materials strength—testing support loads, stiffness, brittleness and other properties—which is essential for many construction, automobile, and medical materials.

How energy gets converted into useful power is the heart of thermodynamics, as well as determining what energy is lost in the process. One specific kind of energy, heat transfer, is crucial in many applications and requires gathering and analyzing temperature data and distributions.

Fluid mechanics, which also has a variety of applications, looks at many properties including pressure drops from fluid flow and aerodynamic drag forces.

Manufacturing is an important step in mechanical engineering. Within the field, researchers investigate the best processes to make manufacturing more efficient. Laboratory methods focus on improving how to measure both thermal and mechanical engineering products and processes. Likewise, machine design develops equipment-scale processes while electrical engineering focuses on circuitry. All this equipment produces vibrations, another field of mechanical engineering, in which researchers study how to predict and control vibrations.

Engineering economics makes mechanical designs relevant and usable in the real world by estimating manufacturing and life cycle costs of materials, designs, and other engineered products.

What skills do mechanical engineers need?

The essence of engineering is problem solving. With this at its core, mechanical engineering also requires applied creativity—a hands on understanding of the work involved—along with strong interpersonal skills like networking, leadership, and conflict management. Creating a product is only part of the equation; knowing how to work with people, ideas, data, and economics fully makes a mechanical engineer.

What tasks do mechanical engineers do?

Careers in mechanical engineering call for a variety of tasks.

- Conceptual design
- Analysis
- Presentations and report writing
- Multidisciplinary teamwork
- Concurrent engineering
- Benchmarking the competition
- Project management

- Prototyping
- Testing
- Measurements
- Data Interpretation
- Developmental design
- Research
- Analysis (FEA and CFD)
- Working with suppliers
- Sales
- Consulting
- Customer service

The future of mechanical engineering

Breakthroughs in materials and analytical tools have opened new frontiers for mechanical engineers. Nanotechnology, biotechnology, composites, computational fluid dynamics (CFD), and acoustical engineering have all expanded the mechanical engineering toolbox.

Nanotechnology allows for the engineering of materials on the smallest of scales. With the ability to design and manufacture down to the elemental level, the possibilities for objects grows immensely. Composites are another area where the manipulation of materials allows for new manufacturing opportunities. By combining materials with different characteristics in innovative ways, the best of each material can be employed and new solutions found. CFD gives mechanical engineers the opportunity to study complex fluid flows analyzed with algorithms. This allows for the modeling of situations that would previously have been impossible. Acoustical engineering examines vibration and sound, providing the opportunity to reduce noise in devices and increase efficiency in everything from biotechnology to architecture.

2

MECHANICS

Mechanics is the area of mathematics and physics concerned with the relationships between force, matter, and motion among physical objects. Forces applied to objects result in displacements, or changes of an object's position relative to its environment.

Theoretical expositions of this branch of physics has its origins in Ancient Greece, for instance, in the writings of Aristotle and Archimedes. During the early modern period, scientists such as Galileo, Kepler, Huygens, and Newton laid the foundation for what is now known as classical mechanics.

As a branch of classical physics, mechanics deals with bodies that are either at rest or are moving with velocities significantly less than the speed of light. It can also be defined as the physical science that deals with the motion of and forces on bodies not in the quantum realm.

ANTIQUITY

The ancient Greek philosophers were among the first to propose that abstract principles govern nature. The main theory of mechanics in antiquity was Aristotelian mechanics, though an alternative theory is exposed in the pseudo-Aristotelian Mechanical Problems, often attributed to one of his successors.

There is another tradition that goes back to the ancient Greeks where mathematics is used more extensively to analyze bodies statically or dynamically, an approach that may have been stimulated by prior work of the Pythagorean Archytas. Examples of this tradition include pseudo-Euclid (On the Balance), Archimedes (On the Equilibrium of Planes, On Floating Bodies), Hero (Mechanica), and Pappus (Collection, Book VIII).

Medieval age

In the Middle Ages, Aristotle's theories were criticized and modified by a number of figures, beginning with John Philoponus in the 6th century. A central problem was that of projectile motion, which was discussed by Hipparchus and Philoponus.

Persian Islamic polymath Ibn Sina published his theory of motion in The Book of Healing (1020). He said that an impetus is imparted to a projectile by the thrower, and viewed it as persistent, requiring external forces such as air resistance to dissipate it. Ibn Sina made distinction between 'force' and 'inclination' (called "mayl"), and argued that an object gained mayl when the object is in opposition to its natural motion. So he concluded that continuation of motion is attributed to the inclination that is transferred to the object, and that object will be in motion until the mayl is spent. He also claimed that a projectile in a vacuum would not stop unless it is acted upon, consistent with Newton's first law of motion.

On the question of a body subject to a constant (uniform) force, the 12th-century Jewish-Arab scholar Hibat Allah Abu'l-Barakat al-Baghdaadi (born Nathanel, Iraqi, of Baghdad) stated that constant force imparts constant acceleration. According to Shlomo Pines, al-Baghdaadi's theory of motion was "the oldest negation of Aristotle's fundamental dynamic law [namely, that a constant force produces a uniform motion], [and is thus an] anticipation in a vague fashion of the fundamental law of classical mechanics [namely, that a force applied continuously produces acceleration]."

Influenced by earlier writers such as Ibn Sina and al-Baghdaadi, the 14th-century French priest Jean Buridan developed the theory of impetus, which later developed into the modern theories of inertia, velocity, acceleration and momentum. This work and others was developed in 14th-century England by the Oxford Calculators such as Thomas Bradwardine, who studied and formulated various laws regarding falling bodies. The concept that the main properties of a body are uniformly accelerated motion (as of falling bodies) was worked out by the 14th-century Oxford Calculators.

Early modern age

Two central figures in the early modern age are Galileo Galilei and Isaac Newton. Galileo's final statement of his mechanics, particularly of falling bodies, is his Two New Sciences (1638). Newton's 1687 Philosophiæ Naturalis Principia Mathematica provided a detailed mathematical account of mechanics, using the newly developed mathematics of calculus and providing the basis of Newtonian mechanics.

There is some dispute over priority of various ideas: Newton's Principia is certainly the seminal work and has been tremendously influential, and many of the mathematics results therein could not have been stated earlier without the development of the calculus. However, many of the ideas, particularly as pertain to inertia and falling bodies, had been developed by prior scholars such as Christiaan Huygens and the less-known medieval predecessors. Precise credit is at times difficult or contentious because scientific language and standards of proof changed, so whether medieval statements are equivalent to modern statements or sufficient proof, or instead similar to modern statements and hypotheses is often debatable.

Modern age

Two main modern developments in mechanics are general relativity of Einstein, and quantum mechanics, both developed in the 20th century based in part on earlier 19th-century ideas. The development in the modern continuum mechanics, particularly in the areas of elasticity, plasticity, fluid dynamics, electrodynamics and thermodynamics of deformable media, started in the second half of the 20th century.

Types of mechanical bodies

The often-used term body needs to stand for a wide assortment of objects, including particles, projectiles, spacecraft, stars, parts of machinery, parts of solids, parts of fluids (gases and liquids), etc.

Other distinctions between the various sub-disciplines of mechanics, concern the nature of the bodies being described. Particles are bodies with little (known) internal structure, treated as mathematical points in classical mechanics. Rigid bodies have size and shape, but retain a simplicity close to that of the particle, adding just a few so-called degrees of freedom, such as orientation in space.

Otherwise, bodies may be semi-rigid, i.e. elastic, or non-rigid, i.e. fluid. These subjects have both classical and quantum divisions of study.

For instance, the motion of a spacecraft, regarding its orbit and attitude (rotation), is described by the relativistic theory of classical mechanics, while the analogous movements of an atomic nucleus are described by quantum mechanics.

Sub-disciplines

The following are two lists of various subjects that are studied in mechanics.

Note that there is also the "theory of fields" which constitutes a separate discipline in physics, formally treated as distinct from mechanics, whether classical fields or quantum fields. But in actual practice, subjects belonging to mechanics and fields are closely interwoven. Thus, for instance, forces that act on particles are frequently derived from fields (electromagnetic or gravitational), and particles generate fields by acting as sources. In fact, in quantum mechanics, particles themselves are fields, as described theoretically by the wave function.

Classical

The following are described as forming classical mechanics:

- Newtonian mechanics, the original theory of motion (kinematics) and forces (dynamics).
- Analytical mechanics is a reformulation of Newtonian mechanics with an emphasis on system energy, rather than on forces. There are two main branches of analytical mechanics:
- Hamiltonian mechanics, a theoretical formalism, based on the principle of conservation of energy.
- Lagrangian mechanics, another theoretical formalism, based on the principle of the least action.
- Classical statistical mechanics generalizes ordinary classical mechanics to consider systems in an unknown state; often used to derive thermodynamic properties.
- Celestial mechanics, the motion of bodies in space: planets, comets, stars, galaxies, etc.
- Astrodynamics, spacecraft navigation, etc.
- Solid mechanics, elasticity, plasticity, viscoelasticity exhibited by deformable solids.

Fracture mechanics

- Acoustics, sound (= density variation propagation) in solids, fluids and gases.
- Statics, semi-rigid bodies in mechanical equilibrium
- Fluid mechanics, the motion of fluids
- Soil mechanics, mechanical behavior of soils
- Continuum mechanics, mechanics of continua (both solid and fluid)
- Hydraulics, mechanical properties of liquids

- Fluid statics, liquids in equilibrium
- Applied mechanics, or Engineering mechanics
- Biomechanics, solids, fluids, etc. in biology
- Biophysics, physical processes in living organisms
- Relativistic or Einsteinian mechanics, universal gravitation.

Quantum

The following are categorized as being part of quantum mechanics:

Schrödinger wave mechanics, used to describe the movements of the wavefunction of a single particle.

Matrix mechanics is an alternative formulation that allows considering systems with a finite-dimensional state space.

Quantum statistical mechanics generalizes ordinary quantum mechanics to consider systems in an unknown state; often used to derive thermodynamic properties.

Particle physics, the motion, structure, and reactions of particles

Nuclear physics, the motion, structure, and reactions of nuclei

Condensed matter physics, quantum gases, solids, liquids, etc.

Historically, classical mechanics had been around for nearly a quarter millennium before quantum mechanics developed. Classical mechanics originated with Isaac Newton's laws of motion in Philosophiæ Naturalis Principia Mathematica, developed over the seventeenth century. Quantum mechanics developed later, over the nineteenth century, precipitated by Planck's postulate and Albert Einstein's explanation of the photoelectric effect. Both fields are commonly held to constitute the most certain knowledge that exists about physical nature.

Classical mechanics has especially often been viewed as a model for other so-called exact sciences. Essential in this respect is the extensive use of mathematics in theories, as well as the decisive role played by experiment in generating and testing them.

Quantum mechanics is of a bigger scope, as it encompasses classical mechanics as a sub-discipline which applies under certain restricted circumstances. According to the correspondence principle, there is no contradiction or conflict between the two subjects, each simply pertains to specific situations. The correspondence principle states that the behavior of systems described by quantum theories reproduces classical physics in the limit of large quantum numbers, i.e. if quantum mechanics is applied to large systems (for e.g. a baseball), the result would almost be the same if classical mechanics had been applied. Quantum mechanics has superseded classical mechanics at the foundation level and is indispensable for the explanation and prediction of processes at the molecular, atomic, and sub-atomic level. However, for macroscopic processes classical mechanics is able to solve problems which are unmanageably difficult (mainly due to computational limits) in quantum mechanics and hence remains useful and well used. Modern descriptions of such behavior begin with a careful definition of such quantities as displacement (distance moved), time, velocity, acceleration, mass, and force. Until about 400 years ago, however, motion was explained from a very different point of view. For example, following the ideas of Greek philosopher and scientist Aristotle, scientists reasoned that a cannonball falls down because its natural position is in the Earth; the sun, the moon, and the stars travel in circles around the earth because it is the nature of heavenly objects to travel in perfect circles.

Often cited as father to modern science, Galileo brought together the ideas of other great thinkers of his time and began to calculate motion in terms of distance travelled from some starting position and the time that it took. He showed that the speed of falling objects increases steadily during the time of their fall. This acceleration is the same for heavy objects as for light ones, provided air friction (air resistance) is discounted. The English mathematician and physicist Isaac Newton improved this analysis by defining force and mass and relating these to acceleration. For objects traveling at speeds close to the speed of light, Newton's laws were superseded by Albert Einstein's theory of relativity. [A sentence illustrating the computational complication of Einstein's theory of relativity.] For atomic and subatomic particles, Newton's laws were superseded by quantum theory. For everyday phenomena, however, Newton's three laws of motion remain the cornerstone of dynamics, which is the study of what causes motion.

RELATIVISTIC

In analogy to the distinction between quantum and classical mechanics, Albert Einstein's general and special theories of relativity have expanded the scope of Newton and Galileo's formulation of mechanics. The differences between relativistic and Newtonian mechanics become significant and even dominant as the velocity of a body approaches the speed of light. For instance, in Newtonian mechanics, the kinetic energy of a free particle is E=1/2mv2, whereas in relativistic mechanics, it is E = (?-1)mc2 (where ? is the Lorentz factor; this formula reduces to the Newtonian expression in the low energy limit).

For high-energy processes, quantum mechanics must be adjusted to account for special relativity; this has led to the development of quantum field theory.

Applied mechanics

Applied mechanics is the branch of science concerned with the motion of any substance that can be experienced or perceived by humans without the help of instruments. In short, when mechanics concepts surpass being theoretical and are applied and executed, general mechanics becomes applied mechanics. It is this stark difference that makes applied mechanics an essential understanding for practical everyday life. It has numerous applications in a wide variety of fields and disciplines, including but not limited to structural engineering, astronomy, oceanography, meteorology, hydraulics, mechanical engineering, aerospace engineering, nanotechnology, structural design, earthquake engineering, fluid dynamics, planetary sciences, and other life sciences. Connecting research between numerous disciplines, applied mechanics plays an important role in both science and engineering.

Pure mechanics describes the response of bodies (solids and fluids) or systems of bodies to external behavior of a body, in either a beginning state of rest or of motion, subjected to the action of forces. Applied mechanics bridges the gap between physical theory and its application to technology.

Composed of two main categories, Applied Mechanics can be split into classical mechanics; the study of the mechanics of macroscopic solids, and fluid mechanics; the study of the mechanics of macroscopic fluids. Each branch of applied mechanics contains subcategories formed through their own subsections as well. Classical mechanics, divided into statics and dynamics, are even further subdivided, with statics' studies split into rigid bodies and rigid structures, and dynamics' studies split into kinematics and kinetics. Like classical mechanics, fluid mechanics is also divided into two sections: statics and dynamics.

Within the practical sciences, applied mechanics is useful in formulating new ideas and theories, discovering and interpreting phenomena, and developing experimental and computational tools. In the application of the natural sciences, mechanics was said to be complemented by thermodynamics, the study of heat and more generally energy, and electromechanics, the study of electricity and magnetism.

Overview

Engineering problems are generally tackled with applied mechanics through the application of theories of classical mechanics and fluid mechanics. Because applied mechanics can be applied in engineering disciplines like civil engineering, mechanical engineering, aerospace engineering, materials engineering, and biomedical engineering, it is sometimes referred to as engineering mechanics.

Science and engineering are interconnected with respect to applied mechanics, as researches in science are linked to research processes in civil, mechanical, aerospace, materials and biomedical engineering disciplines. In civil engineering, applied mechanics' concepts can be applied to structural design and a variety of engineering sub-topics like structural, coastal, geotechnical, construction, and earthquake engineering. In mechanical engineering, it can be applied in mechatronics and robotics, design and drafting, nanotechnology, machine elements, structural analysis, friction stir welding, and acoustical engineering. In aerospace engineering, applied mechanics is used in aerodynamics, aerospace structural mechanics and propulsion, aircraft design and flight mechanics. In materials engineering, applied mechanics' concepts are used in thermoelasticity, elasticity theory, fracture and failure mechanisms, structural design optimisation, fracture and fatigue, active materials and composites, and computational mechanics. Research in applied mechanics can be directly linked to biomedical engineering areas of interest like orthopaedics; biomechanics; human body motion analysis; soft tissue modelling of muscles, tendons, ligaments, and cartilage; biofluid mechanics; and dynamic systems, performance enhancement, and optimal control.

Brief history

The first science with a theoretical foundation based in mathematics was mechanics; the underlying principles of mechanics were first delineated by Isaac Newton in his 1687 book Philosophiæ Naturalis Principia Mathematica. One of the earliest works to define applied mechanics as its own discipline was the three volume Handbuch der Mechanik written by German physicist and engineer Franz Josef Gerstner. The first seminal work on applied mechanics to be published in English was A Manual of Applied Mechanics in 1858 by English mechanical engineer William Rankine. August Föppl, a German mechanical engineer and professor, published Vorlesungen über techische Mechanik in 1898 in which he introduced calculus to the study of applied mechanics.

Applied mechanics was established as a discipline separate from classical mechanics in the early 1920s with the publication of Journal of Applied Mathematics and Mechanics, the creation of the Society of Applied Mathematics and Mechanics, and the first meeting of the International Congress of Applied Mechanics. In 1921 Austrian scientist Richard von Mises started the Journal of Applied Mathematics and Mechanics (Zeitschrift für Angewante Mathematik und Mechanik) and in 1922 with German scientist Ludwig Prandtl founded the Society of Applied Mathematics and Mechanics (Gesellschaft für Angewandte Mathematik und Mechanik). During a 1922 conference on hydrodynamics and aerodynamics in Innsbruck, Austria, Theodore von Kármán, a Hungarian engineer, and Tullio Levi-Civita, an Italian mathematician, met and decided to organize a conference on applied mechanics. In 1924 the first meeting of the International Congress of Applied Mechanics was held in Delft, the Netherlands attended by more than 200 scientist from around the world. Since this first meeting the congress has been held every four years, except during World War II; the name of the meeting was changed to International Congress of Theoretical and Applied Mechanics in 1960.

Due to the unpredictable political landscape in Europe after the First World War and upheaval of World War II many European scientist and engineers emigrated to the United States. Ukrainian engineer Stephan Timoshenko fled the Bolsheviks Red Army in 1918 and eventually emigrated to the U.S. in 1922; over the next twenty-two years he taught applied mechanics at the University of Michigan and Stanford University. Timoshenko authored thirteen textbooks in applied mechanics, many considered the gold standard in their fields; he also founded the Applied Mechanics Division of the American Society of Mechanical Engineers in 1927 and is considered "America's Father of Engineering Mechanics." In 1930 Theodore von Kármán left Germany and became the first director of the Aeronautical Laboratory at the California Institute of Technology; von Kármán would later co-found the Jet Propulsion Laboratory in 1944. With the leadership of Timoshenko and von Kármán, the influx of talent from Europe, and the rapid growth of the aeronautical and defense industries, applied mechanics became a mature discipline in the U.S. by 1950.

Branches

Dynamics

Dynamics, the study of the motion and movement of various objects, can be further divided into two branches, kinematics and kinetics. For classical mechanics, kinematics would be the analysis of moving bodies using time, velocities, displacement, and acceleration. Kinetics would be the study of moving bodies through the lens of the effects of forces and masses. In the context of fluid mechanics, fluid dynamics pertains to the flow and describing of the motion of various fluids.

Statics

The study of statics is the study and describing of bodies at rest. Static analysis in classical mechanics can be broken down into two categories, deformable bodies and non-deformable bodies. When studying deformable bodies, considerations relating to the forces acting on the rigid structures are analyzed. When studying non-deformable bodies, the examination of the structure and material strength is observed. In the context of fluid mechanics, the resting state of the pressure unaffected fluid is taken into account.

NEWTONIAN FOUNDATION

Being one of the first sciences for which a systematic theoretical framework was developed, mechanics was spearheaded by Sir Isaac Newton's "Principia" (published in 1687). It is the "divide and rule" strategy developed by Newton that helped to govern motion and split it into dynamics or statics. Depending on the type of force, type of matter, and the external forces, acting on said matter, will dictate the "Divide and Rule" strategy within dynamic and static studies.

Archimedes' Principle

Archimedes' principle is a major one that contains many defining propositions pertaining to fluid mechanics. As stated by proposition 7 of archimedes' principle, a solid that is heavier than the fluid its placed in, will descend to the bottom of the fluid. If the solid is to be weighed within the fluid, the fluid will be measured as lighter than the weight of the amount of fluid that was displaced by said solid. Further developed upon by proposition 5, if the solid is lighter than the fluid it is placed in, the solid will have to be forcibly immersed to be fully covered by the liquid. The weight of the amount of displaced fluids will then be equal to the weight of the solid.

3

ARISTOTELIAN PHYSICS

Aristotelian physics is the form of natural science described in the works of the Greek philosopher Aristotle (384–322 BC). In his work Physics, Aristotle intended to establish general principles of change that govern all natural bodies, both living and inanimate, celestial and terrestrial – including all motion (change with respect to place), quantitative change (change with respect to size or number), qualitative change, and substantial change ("coming to be" [coming into existence, 'generation'] or "passing away" [no longer existing, 'corruption']). To Aristotle, 'physics' was a broad field that included subjects that would now be called the philosophy of mind, sensory experience, memory, anatomy and biology. It constitutes the foundation of the thought underlying many of his works.

Key concepts of Aristotelian physics include the structuring of the cosmos into concentric spheres, with the Earth at the centre and celestial spheres around it. The terrestrial sphere was made of four elements, namely earth, air, fire, and water, subject to change and decay. The celestial spheres were made of a fifth element, an unchangeable aether. Objects made of these elements have natural motions: those of earth and water tend to fall; those of air and fire, to rise. The speed of such motion depends on their weights and the density of the medium. Aristotle argued that a vacuum could not exist as speeds would become infinite.

Aristotle described four causes or explanations of change as seen on earth: the material, formal, efficient, and final causes of things. As regards living things, Aristotle's biology relied on observation of natural kinds, both the basic kinds and the groups to which these belonged. He did not conduct experiments in the modern sense, but relied on amassing data, observational procedures such as dissection, and making hypotheses about relationships between measurable quantities such as body size and lifespan.

METHODS

A page from an 1837 edition of the ancient Greek philosopher Aristotle's Physica, a book addressing a variety of subjects including the philosophy of nature and topics now part of its modern-day namesake: physics.

—?Aristotle, Physics VIII.1

While consistent with common human experience, Aristotle's principles were not based on controlled, quantitative experiments, so they do not describe our universe in the precise, quantitative way now expected of science. Contemporaries of Aristotle like Aristarchus rejected these principles in favor of heliocentrism, but their ideas were not widely accepted. Aristotle's principles were difficult to disprove merely through casual everyday observation, but later development of the scientific method challenged his views with experiments and careful measurement, using increasingly advanced technology such as the telescope and vacuum pump.

In claiming novelty for their doctrines, those natural philosophers who developed the "new science" of the seventeenth century frequently contrasted "Aristotelian" physics with their own. Physics of the former sort, so they claimed, emphasized the qualitative at the expense of the quantitative, neglected mathematics and its proper role in physics (particularly in the analysis of local motion), and relied on such suspect explanatory principles as final causes and "occult" essences. Yet in his Physics Aristotle characterizes physics or the "science of nature" as pertaining to magnitudes (megethê), motion (or "process" or "gradual change" – kinêsis), and time (chronon) (Phys III.4 202b30–1). Indeed, the Physics is largely concerned with an analysis of motion, particularly local motion, and the other concepts that Aristotle believes are requisite to that analysis.

—?Michael J. White, "Aristotle on the Infinite, Space, and Time" in Blackwell Companion to Aristotle

There are clear differences between modern and Aristotelian physics, the main being the use of mathematics, largely absent in Aristotle. Some recent studies, however, have re-evaluated Aristotle's physics, stressing both its empirical validity and its continuity with modern physics.

Concepts

Peter Apian's 1524 representation of the universe, heavily influenced by Aristotle's ideas. The terrestrial spheres of water and earth (shown in the form of continents and oceans) are at the center of the universe, immediately surrounded by the spheres of air, and then fire, where meteorites and comets were believed to originate. The surrounding celestial spheres from inner to outer are those of the Moon, Mercury, Venus, Sun, Mars, Jupiter, and Saturn, each indicated by a planet symbol. The eighth sphere is the firmament of fixed stars, which include the visible constellations. The precession of the equinoxes caused a gap between the visible and notional divisions of the zodiac, so medieval Christian astronomers created a ninth sphere, the Crystallinum which holds an unchanging version of the zodiac. The tenth sphere is that of the divine prime mover proposed by Aristotle (though each sphere would have an unmoved mover). Above that, Christian theology placed the "Empire of God".

What this diagram does not show is how Aristotle explained the complicated curves that the planets make in the sky. To preserve the principle of perfect circular motion, he proposed that each planet was moved by several nested spheres, with the poles of each connected to the next outermost, but with axes of rotation offset from each other. Though Aristotle left the number of spheres open to empirical determination, he proposed adding to the many-sphere models of previous astronomers, resulting in a total of 44 or 55 celestial spheres.

Elements and spheres

Aristotle divided his universe into "terrestrial spheres" which were "corruptible" and where humans lived, and moving but otherwise unchanging celestial spheres.

Aristotle believed that four classical elements make up everything in the terrestrial spheres: earth, air, fire and water.[a] He also held that the heavens are made of a special weightless and incorruptible (i.e. unchangeable) fifth element called "aether". Aether also has the name "quintessence", meaning, literally, "fifth being".

Aristotle considered heavy matter such as iron and other metals to consist primarily of the element earth, with a smaller amount of the other three terrestrial elements. Other, lighter objects, he believed, have less earth, relative to the other three elements in their composition.

The four classical elements were not invented by Aristotle; they were originated by Empedocles. During the Scientific Revolution, the ancient theory of classical elements was found to be incorrect, and was replaced by the empirically tested concept of chemical elements.

Celestial spheres

According to Aristotle, the Sun, Moon, planets and stars – are embedded in perfectly concentric "crystal spheres" that rotate eternally at fixed rates. Because the celestial spheres are incapable of any change except rotation, the terrestrial sphere of fire must account for the heat, starlight and occasional meteorites. The lowest, lunar sphere is the only celestial sphere that actually comes in contact with the sublunary orb's changeable, terrestrial matter, dragging the rarefied fire and air along underneath as it rotates. Like Homer's æthere (a????) – the "pure air" of Mount Olympus – was the divine counterpart of the air breathed by mortal beings (???, aer). The celestial spheres are composed of the special element aether, eternal and unchanging, the sole capability of which is a uniform circular motion at a given rate (relative to the diurnal motion of the outermost sphere of fixed stars).

The concentric, aetherial, cheek-by-jowl "crystal spheres" that carry the Sun, Moon and stars move eternally with unchanging circular motion. Spheres are embedded within spheres to account for the "wandering stars" (i.e. the planets, which, in comparison with the Sun, Moon and stars, appear to move erratically). Mercury, Venus, Mars, Jupiter, and Saturn are the only planets (including minor planets) which were visible before the invention of the telescope, which is why Neptune and Uranus are not included, nor are any asteroids. Later, the belief that all spheres are concentric was forsaken in favor of Ptolemy's deferent and epicycle model. Aristotle submits to the calculations of astronomers regarding the total number of spheres and various accounts give a number in the neighborhood of fifty spheres. An unmoved mover is assumed for each sphere, including a "prime mover" for the sphere of fixed stars. The unmoved movers do not push the spheres (nor could they, being immaterial and dimensionless) but are the final cause of the spheres' motion, i.e. they explain it in a way that's similar to the explanation "the soul is moved by beauty".

Terrestrial change

Unlike the eternal and unchanging celestial aether, each of the four terrestrial elements are capable of changing into either of the two elements they share a property with: e.g. the cold and wet (water) can transform into the hot and wet (air) or the cold and dry (earth) and any apparent change into the hot and dry (fire) is actually a two-step process. These properties are predicated of an actual substance relative to the work it is able to do; that of heating or chilling and of desiccating or moistening. The four elements exist only with regard to this capacity and relative to some potential work. The celestial element is eternal and unchanging, so only the four terrestrial elements account for "coming to be" and "passing away" – or, in the terms of Aristotle's De Generatione et Corruptione (?e?? ?e??se?? ?a? f?????), "generation" and "corruption".

Natural place

The Aristotelian explanation of gravity is that all bodies move toward their natural place. For the elements earth and water, that place is the center of the (geocentric) universe; the natural place of water is a concentric shell around the earth because earth is heavier; it sinks in water. The natural place of air is likewise a concentric shell surrounding that of water; bubbles rise in water. Finally, the natural place of fire is higher than that of air but below the innermost celestial sphere (carrying the Moon).

In Book Delta of his Physics (IV.5), Aristotle defines topos (place) in terms of two bodies, one of which contains the other: a "place" is where the inner surface of the former (the containing body) touches the outer surface of the other (the contained body). This definition remained dominant until the beginning of the 17th century, even though it had been questioned and debated by philosophers since antiquity. The most significant early critique was made in terms of geometry by the 11th-century Arab polymath al-Hasan Ibn al-Haytham (Alhazen) in his Discourse on Place.

Natural motion

Terrestrial objects rise or fall, to a greater or lesser extent, according to the ratio of the four elements of which they are composed. For example, earth, the heaviest element, and water, fall toward the center of the cosmos; hence the Earth and for the most part its oceans, will have already come to rest there. At the opposite extreme, the lightest elements, air and especially fire, rise up and away from the center.

The elements are not proper substances in Aristotelian theory (or the modern sense of the word). Instead, they are abstractions used to explain the varying natures and behaviors of actual materials in terms of ratios between them.

Motion and change are closely related in Aristotelian physics. Motion, according to Aristotle, involved a change from potentiality to actuality. He gave example of four types of change, namely change in substance, in quality, in quantity and in place.

Aristotle's laws of motion. In Physics he states that objects fall at a speed proportional to their weight and inversely proportional to the density of the fluid they are immersed in. This is a correct approximation for objects in Earth's gravitational field moving in air or water.

Aristotle proposed that the speed at which two identically shaped objects sink or fall is directly proportional to their weights and inversely proportional to the density of the medium through which they move. While describing their terminal velocity, Aristotle must stipulate that there would be no limit at which to compare the speed of atoms falling through a vacuum, (they could move indefinitely fast because there would be no particular place for them to come to rest in the void). Now however it is understood that at any time prior to achieving terminal velocity in a relatively resistance-free medium like air, two such objects are expected to have nearly identical speeds because both are experiencing a force of gravity proportional to their masses and have thus been accelerating at nearly the same rate. This became especially apparent from the eighteenth century when partial vacuum experiments began to be made, but some two hundred years earlier Galileo had already demonstrated that objects of different weights reach the ground in similar times.

UNNATURAL MOTION

Apart from the natural tendency of terrestrial exhalations to rise and objects to fall, unnatural or forced motion from side to side results from the turbulent collision and sliding of the objects as well as transmutation between the elements (On Generation and Corruption).

Chance

In his Physics Aristotle examines accidents (s?µßeß????, symbebekòs) that have no cause but chance. "Nor is there any definite cause for an accident, but only chance (t???, týche), namely an indefinite (????st??, aóriston) cause" (Metaphysics V, 1025a25).

It is obvious that there are principles and causes which are generable and destructible apart from the actual processes of generation and destruction; for if this is not true, everything will be of necessity: that is, if there must necessarily be some cause, other than accidental, of that which is generated and destroyed. Will this be, or not? Yes, if this happens; otherwise not (Metaphysics VI, 1027a29).

Continuum and vacuum

Aristotle argues against the indivisibles of Democritus (which differ considerably from the historical and the modern use of the term "atom"). As a place without anything existing at or within it, Aristotle argued against the possibility of a vacuum or void. Because he believed that the speed of an object's motion is proportional to the force being applied (or, in the case of natural motion, the object's weight) and inversely proportional to the density of the medium, he reasoned that objects moving in a void would move indefinitely fast – and thus any and all objects surrounding the void would immediately fill it. The void, therefore, could never form.

The "voids" of modern-day astronomy (such as the Local Void adjacent to our own galaxy) have the opposite effect: ultimately, bodies off-center are ejected from the void due to the gravity of the material outside.

Four causes

According to Aristotle, there are four ways to explain the aitia or causes of change. He writes that "we do not have knowledge of a thing until we have grasped its why, that is to say, its cause."

Aristotle held that there were four kinds of causes.

Material

The material cause of a thing is that of which it is made. For a table, that might be wood; for a statue, that might be bronze or marble.

"In one way we say that the aition is that out of which. as existing, something comes to be, like the bronze for the statue, the silver for the phial, and their genera" (194b2 3—6). By "genera," Aristotle means more general ways of classifying the matter (e.g. "metal"; "material"); and that will become important. A little later on. he broadens the range of the material cause to include letters (of syllables), fire and the other elements (of physical bodies), parts (of wholes), and even premisses (of conclusions: Aristotle re-iterates this claim, in slightly different terms, in An. Post II. 11).

Formal

The formal cause of a thing is the essential property that makes it the kind of thing it is. In Metaphysics Book ? Aristotle emphasizes that form is closely related to essence and definition. He says for example that the ratio 2:1, and number in general, is the cause of the octave.

"Another [cause] is the form and the exemplar: this is the formula (logos) of the essence (to ti en einai), and its genera, for instance the ratio 2:1 of the octave" (Phys 11.3 194b26—8)... Form is not just shape... We are asking (and this is the connection with essence, particularly in its canonical Aristotelian formulation) what it is to be some thing. And it is a feature of musical harmonics (first noted and wondered at by the Pythagoreans) that intervals of this type do indeed exhibit this ratio in some form in the instruments used to create them (the length of pipes, of strings, etc.). In some sense, the ratio explains what all the intervals have in common, why they turn out the same.

Efficient

The efficient cause of a thing is the primary agency by which its matter took its form. For example, the efficient cause of a baby is a parent of the same species and that of a table is a carpenter, who knows the form of the table. In his Physics II, 194b29—32, Aristotle writes: "there is that which is the primary originator of the change and of its cessation, such as the deliberator who is responsible [sc. for the action] and the father of the child, and in general the producer of the thing produced and the changer of the thing changed".

Aristotle's examples here are instructive: one case of mental and one of physical causation, followed by a perfectly general characterization. But they conceal (or at any rate fail to make patent) a crucial feature of Aristotle's concept of efficient causation, and one which serves to distinguish it from most modern homonyms. For Aristotle, any process requires a constantly operative efficient cause as long as it continues. This commitment appears most starkly to modern eyes in Aristotle's discussion of projectile motion: what keeps the projectile moving after it leaves the hand? "Impetus," "momentum," much less "inertia," are not possible answers. There must be a mover, distinct (at least in some sense) from the thing moved, which is exercising its motive capacity at every moment of the projectile's flight (see Phys VIII. 10 266b29—267a11). Similarly, in every case of animal generation, there is always some thing responsible for the continuity of that generation, although it may do so by way of some intervening instrument (Phys II.3 194b35—195a3).

Final

The final cause is that for the sake of which something takes place, its aim or teleological purpose: for a germinating seed, it is the adult plant, for a ball at the top of a ramp, it is coming to rest at the bottom, for an eye, it is seeing, for a knife, it is cutting.

Goals have an explanatory function: that is a commonplace, at least in the context of action-ascriptions. Less of a commonplace is the view espoused by Aristotle, that finality and purpose are to be found throughout nature, which is for him the realm of those things which contain within themselves principles of movement and rest (i.e. efficient causes); thus it makes sense to attribute purposes not only to natural things themselves, but also to their parts: the parts of a natural whole exist for the sake of the whole. As Aristotle himself notes, "for the sake of" locutions are ambiguous: "A is for the sake of B" may mean that A exists or is undertaken in order to bring B about; or it may mean that A is for B's benefit (An II.4 415b2—3, 20—1); but both types of finality have, he thinks, a crucial role to play in natural, as well as deliberative, contexts. Thus a man may exercise for the sake of his health: and so "health," and not just the hope of achieving it, is the cause of his action (this distinction is not trivial). But the eyelids are for the sake of the eye (to protect it: PA II.1 3) and the eye for the sake of the animal as a whole (to help it function properly: cf. An II.7).

Biology

According to Aristotle, the science of living things proceeds by gathering observations about each natural kind of animal, organizing them into genera and species (the differentiae in History of Animals) and then going on to study the causes (in Parts of Animals and Generation of Animals, his three main biological works).

The four causes of animal generation can be summarized as follows. The mother and father represent the material and efficient causes, respectively. The mother provides the matter out of which the embryo is formed, while the father provides the agency that informs that material and triggers its development. The formal cause is the definition of the animal's substantial being (GA I.1 715a4: ho logos tês ousias). The final cause is the adult form, which is the end for the sake of which development takes place.

ORGANISM AND MECHANISM

The four elements make up the uniform materials such as blood, flesh and bone, which are themselves the matter out of which are created the non-uniform organs of the body (e.g. the heart, liver and hands) "which in turn, as parts, are matter for the functioning body as a whole (PA II. 1 646a 13—24)".

There is a certain obvious conceptual economy about the view that in natural processes naturally constituted things simply seek to realize in full actuality the potentials contained within them (indeed, this is what is for them to be natural); on the other hand, as the detractors of Aristotelianism from the seventeenth century on were not slow to point out, this economy is won at the expense of any serious empirical content.

Mechanism, at least as practiced by Aristotle's contemporaries and predecessors, may have been explanatorily inadequate — but at least it was an attempt at a general account given in reductive terms of the lawlike connections between things. Simply introducing what later reductionists were to scoff at as "occult qualities" does not explain — it merely, in the manner of Molière's famous satirical joke, serves to re-describe the effect. Formal talk, or so it is said, is vacuous.

Things are not however quite as bleak as this. For one thing, there's no point in trying to engage in reductionist science if you don't have the wherewithal, empirical and conceptual, to do so successfully: science shouldn't be simply unsubstantiated speculative metaphysics. But more than that, there is a point to describing the world in such teleologically loaded terms: it makes sense of things in a way that atomist speculations do not. And further, Aristotle's talk of species-forms is not as empty as his opponents would insinuate. He doesn't simply say that things do what they do because that's the sort of thing they do: the whole point of his classificatory biology, most clearly exemplified in PA, is to show what sorts of function go with what, which presuppose which and which are subservient to which. And in this sense, formal or functional biology is susceptible of a type of reductionism. We start, he tells us, with the basic animal kinds which we all pre-theoretically (although not indefeasibly) recognize (cf. PA I.4): but we then go on to show how their parts relate to one another: why it is, for instance, that only blooded creatures have lungs, and how certain structures in one species are analogous or homologous to those in another (such as scales in fish, feathers in birds, hair in mammals). And the answers, for Aristotle, are to be found in the economy of functions, and how they all contribute to the overall well-being (the final cause in this sense) of the animal.

Psychology

According to Aristotle, perception and thought are similar, though not exactly alike in that perception is concerned only with the external objects that are acting on our sense organs at any given time, whereas we can think about anything we choose. Thought is about universal forms, in so far as they have been successfully understood, based on our memory of having encountered instances of those forms directly.

Aristotle's theory of cognition rests on two central pillars: his account of perception and his account of thought. Together, they make up a significant portion of his psychological writings, and his discussion of other mental states depends critically on them. These two activities, moreover, are conceived of in an analogous manner, at least with regard to their most basic forms. Each activity is triggered by its object – each, that is, is about the very thing that brings it about. This simple causal account explains the reliability of cognition: perception and thought are, in effect, transducers, bringing information about the world into our cognitive systems, because, at least in their most basic forms, they are infallibly about the causes that bring them about (An III.4 429a13–18). Other, more complex mental states are far from infallible. But they are still tethered to the world, in so far as they rest on the unambiguous and direct contact perception and thought enjoy with their objects.

Medieval commentary

The Aristotelian theory of motion came under criticism and modification during the Middle Ages. Modifications began with John Philoponus in the 6th century, who partly accepted Aristotle's theory that "continuation of motion depends on continued action of a force" but modified it to include his idea that a hurled body also acquires an inclination (or "motive power") for movement away from whatever caused it to move, an inclination that secures its continued motion. This impressed virtue would be temporary and self-expending, meaning that all motion would tend toward the form of Aristotle's natural motion.

In The Book of Healing (1027), the 11th-century Persian polymath Avicenna developed Philoponean theory into the first coherent alternative to Aristotelian theory. Inclinations in the Avicennan theory of motion were not self-consuming but permanent forces whose effects were dissipated only as a result of external agents such as air resistance, making him "the first to conceive such a permanent type of impressed virtue for non-natural motion". Such a self-motion (mayl) is "almost the opposite of the Aristotelian conception of violent motion of the projectile type, and it is rather reminiscent of the principle of inertia, i.e. Newton's first law of motion."

The eldest Banu Musa brother, Ja'far Muhammad ibn Musa ibn Shakir (800-873), wrote the Astral Motion and The Force of Attraction. The Persian physicist, Ibn al-Haytham (965-1039) discussed the theory of attraction between bodies. It seems that he was aware of the magnitude of acceleration due to gravity and he discovered that the heavenly bodies "were accountable to the laws of physics". During his debate with Avicenna, al-Biruni also criticized the Aristotelian theory of gravity firstly for denying the existence of levity or gravity in the celestial spheres; and, secondly, for its notion of circular motion being an innate property of the heavenly bodies.

Hibat Allah Abu'l-Barakat al-Baghdaadi (1080–1165) wrote al-Mu'tabar, a critique of Aristotelian physics where he negated Aristotle's idea that a constant force produces uniform motion, as he realized that a force applied continuously produces acceleration, a fundamental law of classical mechanics and an early foreshadowing of Newton's second law of motion. Like Newton, he described acceleration as the rate of change of speed.

In the 14th century, Jean Buridan developed the theory of impetus as an alternative to the Aristotelian theory of motion. The theory of impetus was a precursor to the concepts of inertia and momentum in classical mechanics. Buridan and Albert of Saxony also refer to Abu'l-Barakat in explaining that the acceleration of a falling body is a result of its increasing impetus. In the 16th century, Al-Birjandi discussed the possibility of the Earth's rotation and, in his analysis of what might occur if the Earth were rotating, developed a hypothesis similar to Galileo's notion of "circular inertia". He described it in terms of the following observational test:

"The small or large rock will fall to the Earth along the path of a line that is perpendicular to the plane (sath) of the horizon; this is witnessed by experience (tajriba). And this perpendicular is away from the tangent point of the Earth's sphere and the plane of the perceived (hissi) horizon. This point moves with the motion of the Earth and thus there will be no difference in place of fall of the two rocks."

Life and death of Aristotelian physics

The reign of Aristotelian physics, the earliest known speculative theory of physics, lasted almost two millennia. After the work of many pioneers such as Copernicus, Tycho Brahe, Galileo, Descartes and Newton, it became generally accepted that Aristotelian physics was neither correct nor viable. Despite this, it survived as a scholastic pursuit well into the seventeenth century, until universities amended their curricula.

In Europe, Aristotle's theory was first convincingly discredited by Galileo's studies. Using a telescope, Galileo observed that the Moon was not entirely smooth, but had craters and mountains, contradicting the Aristotelian idea of the incorruptibly perfect smooth Moon. Galileo also criticized this notion theoretically; a perfectly smooth Moon would reflect light unevenly like a shiny billiard ball, so that the edges of the moon's disk would have a different brightness than the point where a tangent plane reflects sunlight directly to the eye. A rough moon reflects in all directions equally, leading to a disk of approximately equal brightness which is what is observed. Galileo also observed that Jupiter has moons – i.e. objects revolving around a body other than the Earth – and noted the phases of Venus, which demonstrated that Venus (and, by implication, Mercury) traveled around the Sun, not the Earth.

According to legend, Galileo dropped balls of various densities from the Tower of Pisa and found that lighter and heavier ones fell at almost the same speed. His experiments actually took place using balls rolling down inclined planes, a form of falling sufficiently slow to be measured without advanced instruments.

In a relatively dense medium such as water, a heavier body falls faster than a lighter one. This led Aristotle to speculate that the rate of falling is proportional to the weight and inversely proportional to the density of the medium. From his experience with objects falling in water, he concluded that water is approximately ten times denser than air. By weighing a volume of compressed air, Galileo showed that this overestimates the density of air by a factor of forty. From his experiments with inclined planes, he concluded that if friction is neglected, all bodies fall at the same rate (which is also not true, since not only friction but also density of the medium relative to density of the bodies has to be negligible. Aristotle correctly noticed that medium density is a factor but focused on body weight instead of density. Galileo neglected medium density which led him to correct conclusion for vacuum).

Galileo also advanced a theoretical argument to support his conclusion. He asked if two bodies of different weights and different rates of fall are tied by a string, does the combined system fall faster because it is now more massive, or does the lighter body in its slower fall hold back the heavier body? The only convincing answer is neither: all the systems fall at the same rate.

Followers of Aristotle were aware that the motion of falling bodies was not uniform, but picked up speed with time. Since time is an abstract quantity, the peripatetics postulated that the speed was proportional to the distance. Galileo established experimentally that the speed is proportional to the time, but he also gave a theoretical argument that the speed could not possibly be proportional to the distance.

Further information: Gravity

Modern scholars differ in their opinions of whether Aristotle's physics were sufficiently based on empirical observations to qualify as science, or else whether they were derived primarily from philosophical speculation and thus fail to satisfy the scientific method.

Carlo Rovelli has argued that Aristotle's physics are an accurate and non-intuitive representation of a particular domain (motion in fluids), and thus are just as scientific as Newton's laws of motion, which also are accurate in some domains while failing in others (i.e. special and general relativity).

4

THERMODYNAMIC FUNCTIONS

The fundamental thermodynamic equations follow from five primary thermodynamic definitions and describe internal energy, enthalpy, Helmholtz energy, and Gibbs energy in terms of their natural variables. Here they will be presented in their differential forms. The fundamental thermodynamic equations describe the thermodynamic quantities U, H, G, and A in terms of their natural variables. The term "natural variable" simply denotes a variable that is one of the convenient variables to describe U, H, G, or A. When considered as a whole, the four fundamental equations demonstrate how four important thermodynamic quantities depend on variables that can be controlled and measured experimentally. Thus, they are essentially equations of state, and using the fundamental equations, experimental data can be used to determine sought-after quantities like G or H.

TEMPERATURE

Thermodynamic temperature is the measure of absolute temperature and is one of the principal parameters of thermodynamics. A thermodynamic temperature reading of zero denotes the point at which the fundamental physical property that imbues matter with a temperature, transferable kinetic energy due to atomic motion, begins. In science, thermodynamic temperature is measured on the Kelvin scale and the unit of measure is the kelvin (unit symbol: K). For comparison, a temperature of 295 K is a comfortable one, equal to 21.85 °C and 71.33 °F.

At the zero point of thermodynamic temperature, absolute zero, the particle constituents of matter have minimal motion and can become no colder. Absolute zero, which is a temperature of zero kelvin (0 K), is precisely equal to "273.15 °C and "459.67 °F. Matter at absolute zero has no remaining transferable average kinetic energy and the only remaining particle motion is due to an ever-pervasive quantum mechanical phenomenon called zero-point energy.

Though the atoms in, for instance, a container of liquid helium that was precisely at absolute zero would still jostle slightly due to zero-point energy, a theoretically perfect heat engine with such helium as one of its working fluids could never transfer any net kinetic energy (heat energy) to the other working fluid and no thermodynamic work could occur.

Temperature is generally expressed in absolute terms when scientifically examining temperature's interrelationships with certain other physical properties of matter such as its volume or pressure, or the wavelength of its emitted black-body radiation. Absolute temperature is also useful when calculating chemical reaction rates. Furthermore, absolute temperature is typically used in cryogenics and related phenomena like superconductivity, as per the following example usage: "Conveniently, tantalum's transition temperature (Tc) of 4.4924 kelvin is slightly above the 4.2221 K boiling point of helium."

The Kelvin scale is also used in everyday life—often without people realizing it—due to color film photography and the need for films that were balanced for two kinds of tungsten-filament studio lights as well as noonday sun. Since the temperature of objects that emit black-body radiation, like the sun and tungsten filaments, had long been measured in kelvins, the color temperature of photographic lighting—and even regular LED bulbs for room lighting nowadays—are measured in kelvins, as per the following example usage: "Photographers who don't own a color temperature meter may illuminate their scenes with ordinary 3200 kelvin (warm-white) LED bulbs and set their digital cameras' white point to 'tungsten.'"

ADIABATIC PROCESSES

In thermodynamics, an adiabatic process is a type of thermodynamic process which occurs without transferring heat or mass between the system and its surroundings. Unlike an isothermal process, an adiabatic process transfers energy to the surroundings only as work. It also conceptually supports the theory used to explain the first law of thermodynamics and is therefore a key thermodynamic concept.

Some chemical and physical processes occur too rapidly for energy to enter or leave the system as heat, allowing a convenient «adiabatic approximation». For example, the adiabatic flame temperature uses this approximation to calculate the upper limit of flame temperature by assuming combustion loses no heat to its surroundings.

In meteorology and oceanography, adiabatic cooling produces condensation of moisture or salinity, oversaturating the parcel. Therefore, the excess must be removed. There, the process becomes a pseudo-adiabatic process whereby the liquid water or salt that condenses is assumed to be removed upon formation by idealized instantaneous precipitation. The pseudoadiabatic process is only defined for expansion because a compressed parcel becomes warmer and remains undersaturated.

A process without transfer of heat to or from a system, so that Q = 0, is called adiabatic, and such a system is said to be adiabatically isolated. The assumption that a process is adiabatic is a frequently made simplifying assumption. For example, the compression of a gas within a cylinder of an engine is assumed to occur so rapidly that on the time scale of the compression process, little of the system's energy can be transferred out as heat to the surroundings. Even though the cylinders are not insulated and are quite conductive, that process is idealized to be adiabatic. The same can be said to be true for the expansion process of such a system.

The assumption of adiabatic isolation is useful and often combined with other such idealizations to calculate a good first approximation of a system›s behaviour. For example, according to Laplace, when sound travels in a gas, there is no time for heat conduction in the medium, and so the propagation of sound is adiabatic. For such an adiabatic process, the modulus of elasticity (Young›s modulus) can be expressed as E = ^{3}P, where ^{3} is the ratio of specific heats at constant pressure and at constant volume (^{3} = Cp/Cv) and P is the pressure of the gas.

An adiabatic process is one in which no heat is gained or lost by the system. The first law of thermodynamics with Q=0 shows that all the change in internal energy is in the form of work done. This puts a constraint on the heat engine process leading to the adiabatic condition shown below. This condition can be used to derive the expression for the work done during an adiabatic process.

The ratio of the specific heats ^{3} = CP/CV is a factor in determining the speed of sound in a gas and other adiabatic processes as well as this application to heat engines. This ratio ^{3} = 1.66 for an ideal monoatomic gas and ^{3} = 1.4 for air, which is predominantly a diatomic gas.

Various Applications of the Adiabatic Assumption

For a closed system, one may write the first law of thermodynamics as : ”U = Q – W, where ”U denotes the change of the system's internal energy, Q the quantity of energy added to it as heat, and W the work done by the system on its surroundings.

If the system has such rigid walls that work cannot be transferred in or out (W = 0), and the walls are not adiabatic and energy is added in the form of heat (Q > 0), and there is no phase change, then the temperature of the system will rise.

If the system has such rigid walls that pressure–volume work cannot be done, but the walls are adiabatic (Q = 0), and energy is added as isochoric work in the form of friction or the stirring of a viscous fluid within the system (W < 0), and there is no phase change, then the temperature of the system will rise.

If the system walls are adiabatic (Q = 0) but not rigid (W “‘ 0), and, in a fictive idealized process, energy is added to the system in the form of frictionless, non-viscous pressure–volume work (W < 0), and there is no phase change, then the temperature of the system will rise. Such a process is called an isentropic process and is said to be "reversible". Ideally, if the process were reversed the energy could be recovered entirely as work done by the system. If the system contains a compressible gas and is reduced in volume, the uncertainty of the position of the gas is reduced, and seemingly would reduce the entropy of the system, but the temperature of the system will rise as the process is isentropic (“S = 0). Should the work be added in such a way that friction or viscous forces are operating within the system, then the process is not isentropic, and if there is no phase change, then the temperature of the system will rise, the process is said to be "irreversible", and the work added to the system is not entirely recoverable in the form of work.

If the walls of a system are not adiabatic, and energy is transferred in as heat, entropy is transferred into the system with the heat. Such a process is neither adiabatic nor isentropic, having Q > 0, and ”S > 0 according to the second law of thermodynamics.

Naturally occurring adiabatic processes are irreversible (entropy is produced).

The transfer of energy as work into an adiabatically isolated system can be imagined as being of two idealized extreme kinds. In one such kind, no entropy is produced within the system (no friction, viscous dissipation, etc.), and the work is only pressure-volume work (denoted by P dV). In nature, this ideal kind occurs only approximately because it demands an infinitely slow process and no sources of dissipation.

The other extreme kind of work is isochoric work (dV = 0), for which energy is added as work solely through friction or viscous dissipation within the system. A stirrer that transfers energy to a viscous fluid of an adiabatically isolated system with rigid walls, without phase change, will cause a rise in temperature of the fluid, but that work is not recoverable. Isochoric work is irreversible. The second law of thermodynamics observes that a natural process, of transfer of energy as work, always consists at least of isochoric work and often both of these extreme kinds of work. Every natural process, adiabatic or not, is irreversible, with ”S > 0, as friction or viscosity are always present to some extent.

THERMODYNAMIC IDENTITY

A useful summary relationship called the thermodynamic identity makes use of the power of calculus and particularly partial derivatives. It may be applied to examine processes in which one or more state variables is held constant, e.g., constant volume, constant pressure, etc. The thermodynamic identity holds true for any infinitesmal change in a system so long at the pressure and temperature are well defined. It is presumed that the number of particles is constant (i.e., you are dealing with the same system before and after the change).

Thermodynamic identity:

$$dU = TdS - PdV$$

U = internal energy

S = entropy

V = volume

T = temperature

P = pressure

“d” denotes the total differential of the associated quantity

Often the definition of temperature is made in terms of the average translational kinetic energy of the particles; this is called the kinetic temperature. An alternative defintion of temperature can be made from the thermodynamic identity:

If we hold the volume constant, this leads to an expression for temperature as a partial derivative of the entropy with respect to the internal energy.

This definition implies that you are holding both the volume and the number of particles constant in taking the derivative. This can be applied to the expression for the entropy of a monoatomic ideal gas:

Taking the partial derivative of this with respect to U to get the temperature requires some gymnastics with the logarithms. Using the rules for the logarithms of products allows us to express the logarithm as just ln(U3/2) plus some other log terms which do not contain U and therefore will drop out when the derivative is taken. Then remember that ln(U)3/2=3/2ln(U) and that the derivative of ln(U) is just 1/U. This finally gives us:

This relationship for the internal energy is just that which is obtained from equipartition of energy. Since the internal energy of a monoatomic gas is just the translational kinetic energy of the molecules, this is in agreement with the kinetic temperature mentioned above, so the two definitions of temperature are equivalent for this case.

LAWS OF THERMODYNAMICS

The laws of thermodynamics define a group of physical quantities, such temperature, energy,and entropy, that characterize thermodynamic syste ms in thermo dynamic equilibrium. The laws also use various parameters for thermodynamic processes, such as thermodynamic work and heat, and establish relationships between them. They state empirical facts that form a basis of precluding the possibility of certain phenomena, such as perpetual motion. In addition to their use in thermodynamics, they are important fundamental laws of physics in general, and are applicable in other natural sciences.

Traditionally, thermodynamics has recognized three fundamental laws, simply named by an ordinal identification, the first law, the second law, and the third law.A more fundamental statement was later labelled as the zeroth law, after the first three laws had been established.

The zeroth law of thermodynamics defines thermal equilibrium and forms a basis for the definition of temperature: If two systems are each in thermal equilibrium with a third system, then they are in thermal equilibrium with each other.

The first law of thermodynamics states that, when energy passes into or out of a system (as work, heat, or matter), the system's internal energy changes in accord with the law of conservation of energy.

The second law of thermodynamics states that in a natural thermodynamic process, the sum of the entropies of the interacting thermodynamic systems never decreases. Another form of the statement is that heat does not spontaneously pass from a colder body to a warmer body.

The third law of thermodynamics states that a system's entropy approaches a constant value as the temperature approaches absolute zero. With the exception of non-crystalline solids (glasses) the entropy of a system at absolute zero is typically close to zero.

The first and second law prohibit two kinds of perpetual motion machines, respectively: the perpetual motion machine of the first kind which produces work with no energy input, and the perpetual motion machine of the second kind which spontaneously converts thermal energy into mechanical work.**FIRST LAW OF THERMODYNAMICS**

The first law of thermodynamics, also known as Law of Conservation of Energy, states that energy can neither be created nor destroyed; energy can only be transferred or changed from one form to another. For example, turning on a light would seem to produce energy; however, it is electrical energy that is converted.

A way of expressing the first law of thermodynamics is that any change in the internal energy ("E) of a system is given by the sum of the heat (q) that flows across its boundaries and the work (w) done on the system by the surroundings:

[latex]\Delta E = q + w[/latex]

This law says that there are two kinds of processes, heat and work, that can lead to a change in the internal energy of a system. Since both heat and work can be measured and quantified, this is the same as saying that any change in the energy of a system must result in a corresponding change in the energy of the surroundings outside the system. In other words, energy cannot be created or destroyed. If heat flows into a system or the surroundings do work on it, the internal energy increases and the sign of q and w are positive. Conversely, heat flow out of the system or work done by the system (on the surroundings) will be at the expense of the internal energy, and q and w will therefore be negative.

SECOND LAW OF THERMODYNAMICS

The second law of thermodynamics says that the entropy of any isolated system always increases. Isolated systems spontaneously evolve towards thermal equilibrium—the state of maximum entropy of the system. More simply put: the entropy of the universe (the ultimate isolated system) only increases and never decreases.

A simple way to think of the second law of thermodynamics is that a room, if not cleaned and tidied, will invariably become more messy and disorderly with time – regardless of how careful one is to keep it clean. When the room is cleaned, its entropy decreases, but the effort to clean it has resulted in an increase in entropy outside the room that exceeds the entropy lost.

Thermodynamic Potentials, Maxwell Relations

In this chapter we shall introduce a number of equivalent thermodynamic representations. We shall derive transformations between them and prove that all these representations are, in fact, equivalent.

Equilibrium of a system is either characterized by an entropy maximum or equivalently by an energy minimum. These two equivalent principles can be summarized as follows.

Entropy Maximum Principle: Any unconstrained internal parameter assumes an equilibrium value such that the entropy of the system is maximized for the given value of total internal energy.

Energy Minimum Principle: Any unconstrained internal parameter assumes an equilibrium value such that the energy of the system is minimized for the given value of total entropy.

Let's consider a cylinder that has two compartments separated by a wall as shown in the figure. The constraints are removed and the gas expands irreversibly but adiabatically. Thus, the total internal energy is conserved and the total entropy increases.

In the next figure the same system undergoes a reversible and adiabatic transformation. An external force enforces a reversible motion of the wall. Work is done on the surroundings but no heat is transferred to the system. Thus, the total entropy is conserved and the total internal energy drops.

These examples illustrate that no matter how the final equilibrium state is brought about, it satisfies both extremal conditions.

We now illustrate this more formally. In the next figure the configuration space of the same thermodynamics system is shown. U and S correspond to the internal energy and total entropy, respectively. Xj(1) corresponds to a particular extensive parameter of one of the volumes. For clarity, other parameters are not shown. The curvature of the hypersurface is defined by the postulate dS / dU > 0 and by the fact that S is a homogeneous function of U. The entropy maximum principle for the composite is shown, thus the system evolves along the intersection of the hypersurface with the plane of constant U.

Even though we discussed the equivalence of both principles and understand it intuitively we have not yet proven it. We assume the entropy maximum principle, that is

and

For convenience, let›s denote the first derivative by A

Since A equals zero, U has an extremum. To decide whether this extremum is a maximum, minimum or a saddle point we study the sign of the second derivative.

The second derivative is

Thus for A = 0 we get

Inserting we obtain.

Since holds we get

thus U has a minimum. q.e.d.

Maxwell Relations

Elementary theorems of calculus state that partial derivatives of a function f can be exchanged if the original function fulfills certain criteria. In general, these criteria are that f is differentiable and that its derivative fx is differentiable. These criteria are met by the entropic fundamental equation, which was found to be homogeneous first-order in its extensive parameters. Since all representations of the fundamental equations are equivalent we can assume that all behave well. Thus,

is a valid equation. For example, since

we obtain

Thus from follows

This relation is one of many relations between partial derivatives of state functions. These relations are called Maxwell relations. They are not necessary for the formal structure of thermodynamics but do introduce substantial simplifications in thermodynamics calculations. In general, for systems with t+1 natural variables there are t(t+1)/2 mixed pairs of second derivatives for each fundamental equation, i.e., there are t(t+1)/2 Maxwell relations. For instance, a system with, let's say, 2 degrees of freedom has 3 natural and 2 independent variables. Thus t=3 implies 3 Maxwell relations per potential. There are 23-1 potential functions, and consequently 21 Maxwell relations. Let's instead list four of the most useful Maxwell relations, derived from the four potentials.

Each side has a thermodynamic potential; each corner has a thermodynamic variable. The extensive variables are on the right; the intensive variables are on the left side. Each potential of flanked by its natural coordinates. E.g., the enthalpy H is a function of S and p. You can double-check all other thermodynamics potentials by comparing to the equations. This diagram highlights nicely the conjugate thermodynamics variable. Since all potentials have the dimension of energy a pair of conjugated variables also is energy. Conjugated are connected by arrows. The diagram is used as follows:

Again, each potential is flanked by its natural coordinates.

Adding conjugated variable pairs derives the fundamental equations. An arrow beginning at the natural variable of a potential is indicating a plus in the sum. For instance, H = +ST +pV, but G = +pV -TS.

Derivation of Maxwell relations by using the corners of the Koenig›s diagram.

Now we have a set of tools available and we can derive several quantities and equations that characterize many thermodynamics systems and that are experimentally accessible. This is an important step because often thermodynamics relations contain quantities such as the thermodynamic potentials, the entropy or their derivatives. These quantities are usually not easily known. Therefore, being able to transform fundamental equations in such a way that experimentally know variables are employed allows us to actually apply thermodynamics to real problems.

Quantities that are of particular usefulness are:

The molar heat capacity at constant volume corresponding to equation

The molar heat capacity at constant pressure corresponding to equation

The isobaric expansion coefficient

The isothermal compressibility

These parameters are mutually dependent, as the following considerations will show. Does the heat capacity cp change as a function of the pressure? Taking the derivative of.

The differentiation can be exchanged according to general calculus. Thus

The last step involved the (inverse) Maxwell equation. We obtain

Here we see how very useful the Maxwell relations are. We could use them to get rid of the entropy in the equation. After all, we would have no clue how to calculate the entropy without having some fundamental equation describing some system. In an analog way we can find

Inserting the ideal gas equation of state pV=RT into yields

Let's find a general expression for the difference of the heat capacities at constant pressure and constant volume.

Inserting into yields

Applying Maxwell relation we get

Inserting the isobaric expansion coefficient and the isothermal compressibility we get

For an ideal gas pV=RT. Inserting this into and yields

$$C_p - C_v = R$$

HEAT CAPACITY AND EQUATION OF STATE

As shown originally by Count Rumford, there is an equivalence between heat (measured in calories) and mechanical work (measured in joules) with a definite conversion factor between the two. The conversion factor, known as the mechanical equivalent of heat, is 1 calorie = 4.184 joules. (There are several slightly different definitions in use for the calorie. The calorie used by nutritionists is actually a kilocalorie.) In order to have a consistent set of units, both heat and work will be expressed in the same units of joules.

The amount of heat that a substance absorbs is connected to its temperature change via its molar specific heat c, defined to be the amount of heat required to change the temperature of 1 mole of the substance by 1 K. In other words, c is the constant of proportionality relating the heat absorbed (d2 Q) to the temperature change (dT) according to d2 Q = nc dT, where n is the number of moles. For example, it takes approximately 1 calorie of heat to increase the temperature of 1 gram of water by 1 K. Since there are 18 grams of water in 1 mole, the molar heat capacity of water is 18 calories per K, or about 75 joules per K. The total heat capacity C for n moles is defined by C = nc.

However, since d2 Q is not an exact differential, the heat absorbed is path-dependent and the path must be specified, especially for gases where the thermal expansion is significant. Two common ways of specifying the path are either the constant-pressure path or the constant-volume path. The two different kinds of specific heat are called cP and cV respectively, where the subscript denotes the quantity that is being held constant. It should not be surprising that cP is always greater than cV, because the substance must do work against the surrounding atmosphere as it expands upon heating at constant pressure but not at constant volume. In fact, this difference was used by the 19th-century German physicist Julius Robert von Mayer to estimate the mechanical equivalent of heat.

JACOBIAN METHOD

In numerical linear algebra, the Jacobi method is an iterative algorithm for determining the solutions of a strictly diagonally dominant system of linear equations. Each diagonal element is solved for, and an approximate value is plugged in. The process is then iterated until it converges. This algorithm is a stripped-down version of the Jacobi transformation method of matrix diagonalization. The method is named after Carl Gustav Jacob Jacobi.

The Jacobi method is easily derived by examining each of the n equations in the linear system Ax =b in isolation. If in the i th equation we solve for the value of xi while assuming the other entries of x remain fixed, we obtain

This suggests an iterative method defined by

which is the Jacobi method. Note that the order in which the equations are examined is irrelevant, since the Jacobi method treats them independently. For this reason, the Jacobi method is also known as the method of simultaneous displacements, since the updates could in principle be done simultaneously.

Simultaneous displacements, method of Jacobi method.

In matrix terms, the definition of the Jacobi method in can be expressed as

$$x^{(h)} = D^{-1} (L+U)x^{(h-1)}+D^{-1}b_1$$

where the matrices D-L, and -V represent the diagonal, the strictly lower-triangular, and the strictly upper-triangular parts of A, respectively.

The pseudocode for the Jacobi method is given in Figure. Note that an auxiliary storage vector,‘x is used in the algorithm. It is not possible to update the vector x in place, since values from x(k-1) are needed throughout the computation of x(h).

JOULE-THOMSON PROCESS

The Joule Thomson effect refers to a thermodynamic process that occurs when the expansion of fluid takes place from high pressure to low pressure at constant enthalpy. Furthermore, the approximation of such process takes place in the real world by facilitating an expansion of fluid from high pressure to low pressure across a valve. Also, the Joule-Thomson effect and inversion temperature is dependent on the gas's pressure before expansion.

Introduction to Joule Thomson Effect

The Joule-Thomson effect, also known as the Joule-Kelvin effect, refers to the change which takes place in fluid's temperature as it flows from a region of higher pressure to lower pressure. One can describe the Joule-Thomson effect by means of the Joule-Thomson coefficient. Also, the Joule-Thomson coefficient is the partial pressure derivative with respect to temperature at constant enthalpy.

The Joule-Thomson coefficient would vary as a function of temperature and pressure. Furthermore, the Joule-Thomson coefficient would vary from one fluid to another. Also, the inversion curve refers to the curve that is described by the Joule-Thomson coefficient equalling zero as a function of temperature and pressure.

The adiabatic (no h-eat exchanged) expansion of a gas can take place in a number of ways, as part of the Joule-Thomson effect experiment. The change in temperature which a gas experiences during expansion is dependent not only on the initial pressure and final pressure but also on the manner the expansion takes place.

If the expansion process is reversible, such that the gas would be always in thermodynamic equilibrium, the expansion is isentropic. Here, the temperature of the gas decreases and it does positive work during the expansion.

The gas does no work in free expansion and no absorption of heat takes place. As such, there is a conservation of internal energy. Here, the ideal gas's temperature would remain constant but the real gas's temperature would decrease, except at extremely high temperature.

The Joule–Thomson expansion refers to a method of expansion in which a gas or liquid at pressure P1, without a considerable change in kinetic energy, flows into a region of lower pressure P2. The expansion is certainly inherently irreversible.

During such an expansion, there would be no change in enthalpy. Furthermore, the determination of whether the internal energy increases or decreases takes place by whether work is done by the liquid or on the fluid. This means that it is determined by the initial and final states of the fluid's properties.

The Joule–Thomson coefficient makes possible the quantification of the temperature change during a Joule–Thomson expansion display. Furthermore, this coefficient may be either positive or negative. Moreover, being positive corresponds to cooling while being negative corresponds to heating.

The coefficient turns out to be negative at extremely high and extremely low temperatures. Furthermore, it is negative at very high pressure at all temperatures. Also, the maximum inversion temperature (621 K for N2) takes place as the zero pressure approaches.

At temperatures that are below the gas-liquid coexistence curve, the condensation of N2 takes place to form a liquid. Moreover, the coefficient would once again become negative. Thus, for N2 gas below 621 K, a Joule–Thomson expansion can be used to cool the gas until the formation of liquid N2 takes place.

11. CONDITION FOR EQUILIBRIUM AND STABILITY IN AN ISOLATED SYSTEM

Thermodynamic methods make it possible to show that if a system is in the state of thermodynamic stability, the following relationships must be satisfied for any substance:

$$C_v > 0$$

and

i.e. in the first place, the isochoric (constant-volume) heat capacity cv is always positive and, in the second place, in an isothermal process an increase in pressure always results in a decrease in the volume of substance. The condition is called the condition for thermal stability, and the condition is referred to as the condition for mechanical stability.

The conditions and can be elucidated by the so-called Le Chatelier principle, which states that:

If a system is subjected to a constraint whereby the equilibrium is modified, a change takes place, if possible, which partially annuls the constraint.

These conditions for thermodynamic stability of a system are clear and require no formal derivation. Imagine the heat capacity cv of some substance to be negative. This would mean that, since isochoric addition of heat would result not in an increase but in a decrease in temperature. This would mean that the more heat added to the substance undergoing an isochoric process, the greater the difference between the temperatures of the substance and the heat source (surroundings). As a result of the increase in the difference between the temperatures of the substance and surroundings, the whole system (i.e. the substance and the heat source) would continue to deviate from the state of equilibrium instead of tending to reach it, with the process developing at an ever increasing rate.

Thus, the system would be unstable: even a negligible difference in the temperatures of the substance and the surroundings would cause an avalanche-type increase in the system's lability. On the other hand, similar reasoning for the case when cv > 0 brings us to the logical conclusion that heat transfer between the substance considered and the surroundings, accompanied by an increase in substance temperature, will cease when the temperatures of the surroundings and substance become equal, and a state of equilibrium establishes in the system.

The validity of the condition can be assured as follows. Assume that for a substance which means that with an increase in volume the pressure in the substance will rise. Just as before, consider a system consisting of two components, the substance and the surroundings; there is heat transfer between the substance and the surroundings, in the course of which the temperatures of the two constituents are the same. Let the pressure of the substance increase in respect to the pressure of the surroundings by an infinitesimal value. This will clearly result in a certain expansion of the substance, and the volume of the surroundings will decrease (they will contract). However, in accordance with condition, this will cause a further increase in the pressure of the substance considered, which, in turn, will be accompanied by an increase in the volume of the substance, etc. Proceeding at an ever increasing rate, the process will lead to a limitless expansion of the substance at an infinitely large increase in the substance pressure.

If we consider another case when the initial pressure of the substance is somewhat lower than the pressure of the surroundings, similar reasonings will lead us to the conclusion that an avalanche-type decrease in the volume of the substance with a drop in substance pressure is inevitable.

Thus, in both cases the system considered will be unstable.

On the other hand, if

and if the pressure of the substance investigated exceeds the pressure of the surroundings, an expansion of the substance will lead to a decrease in its pressure until the pressure of the substance and that of the surroundings become equal, i.e. until the system reaches a state of equilibrium.

Let us proceed now to consider the conditions for the equilibrium in thermodynamic systems.

Of the variety of thermodynamic systems differing from each other in the various modes of interactions (conjugation) with the surroundings, of the greatest practical interest are the conditions for equilibrium in an isolated thermodynamic system.

Consider such an isolated system, illustrated in Figure below. Imagine this system divided into two parts 1 and 2 and find the conditions for equilibrium between the two subsystems.

At the same time consider an infinitesimal process proceeding inside the isolated system such that it causes either a change in the volume of each of the subsystems or a change in the internal energy of the subsystems or a change in both volume and internal energy. Let the volume of the first subsystem change by dV1, and the internal energy by dU1 and the volume and internal energy of the second substance by dV2 and dU2, respectively. Inasmuch as the volume and the internal energy of the entire system remain constant, dV1 = - dV2 and dU1= - dU2; in other words, the change in the volume (or internal energy) of the first subsystem is equal to the change in the volume (or internal energy) of the second subsystem.

Previously, an important equilibrium criterion was established for an isolated system: the entropy of an isolated system in thermodynamic equilibrium was shown to preserve a constant (maximum) value, i.e. in a state of equilibrium

$$dS_{sys} = 0.$$

Since entropy is an additive quantity, for the case under consideration,

$$S_{sys} = S_1 + S_2,$$

and in accordance with Equation,

$$dS_{sys} = dS_1 + dS_2 = 0.$$

From equation

$$Tas = dU + pdV$$

it follows that

Thus, for subsystem 1,

and for subsystem 2,

In accordance with Eq. (5.68), we get:

Equation can be presented in the following form:

It was mentioned above that the volume and internal energy of each of the subsystems can change independently from each other, i.e. a process is possible in the course of which the volume of each subsystem changes and their internal energies remain unchanged and, vice versa, a change in the internal energy of the subsystems may not cause a change in their volumes. In other words, the differentials dV1 and dU1 are independent, in principle. Then, for the left-hand side of Equation to be equal to zero, the coefficients of the differentials dV1 and dU1, present in this equation, should be independently equal to zero, i.e. it is required that

and

From we obtain:

$$T_1 = T_2$$

and from equation, allowing for equation, we get:

$$p_1 = p_2$$

We shall, obviously, come to the same conclusion, irrespective of the number of subsystems into which the system is imagined to be divided. Thus, we arrived at the conclusion that in equilibrium the temperature and pressure are the same in all parts of an isolated system.

The question arises of whether the obtained conclusion is valid for any isolated system or whether in deriving it some simplifying assumptions, restricting the sphere of application of this conclusion, were made.

In fact, a number of restrictions was admitted. Firstly, the combined mathematical statement of the first and second laws of thermodynamics was used not in its most general form

$$TdS = dU + dL$$

but in the form

$$TdS = dU + pdV,$$

i.e. we restricted ourselves to considering the case when the only kind of work done is the work of expansion. If we considered other kinds of work (for instance, a system placed in a potential field; an example is a gas in a gravitational field), we would have other conditions for the equilibrium of a system. It is also easy to find that condition, requiring that the temperature be the same over the entire volume of a system, would remain unchanged, and only condition would change. For instance, for a gas placed in a gravitational field it would follow that the pressure in the column of the gas would increase with diminishing height.

Secondly, we assumed that there exist no peculiarities for the interface separating the two subsystems which should be taken into account. This assumption is not valid when a substance is in different phases in the subsystems; strictly speaking, account should be taken of the surface layer, which, as it will be seen below, has a number of special properties. One more member must then be added, and that is the energy of the surface layer. It will be noted, however, that condition will then remain unchanged. As regards condition, for the case with a flat interface this condition will also remain unchanged; but if the interface is a curved surface, condition will be replaced by another one.

Thermodynamic Inequalities

Growing economic inequalities are observed in several countries throughout the world. Following Pareto, the power-law structure of these inequalities has been the subject of much theoretical and empirical work. But their nonequilibrium dynamics, e.g. after a policy change, remains incompletely understood. Here we introduce a thermodynamical theory of inequalities based on the analogy between economic stratification and statistical entropy. Within this framework we identify the combination of upward mobility with precariousness as a fundamental driver of inequality. We formalize this statement by a "second-law" inequality displaying upward mobility and precariousness as thermodynamic conjugate variables. We estimate the time scale for the "relaxation" of the wealth distribution after a sudden change of the after-tax return on capital. Our method can be generalized to gain insight into the dynamics of inequalities in any Markovian model of socioeconomic interactions.

Third Law of Thermodynamics

The third law of thermodynamics states as follows, regarding the properties of closed systems in thermodynamic equilibrium:

The entropy of a system approaches a constant value as its temperature approaches absolute zero. This constant value cannot depend on any other parameters characterizing the closed system, such as pressure or applied magnetic field. At absolute zero (zero kelvins) the system must be in a state with the minimum possible energy. Entropy is related to the number of accessible microstates, and there is typically one unique state (called the ground state) with minimum energy. In such a case, the entropy at absolute zero will be exactly zero. If the system does not have a well-defined order (if its order is glassy, for example), then there may remain some finite entropy as the system is brought to very low temperatures, either because the system becomes locked into a configuration with non-minimal energy or because the minimum energy state is non-unique.

The constant value is called the residual entropy of the system. The entropy is essentially a state-function meaning the inherent value of different atoms, molecules, and other configurations of particles including subatomic or atomic material is defined by entropy, which can be discovered near 0 K. The Nernst–Simon statement of the third law of thermodynamics concerns thermodynamic processes at a fixed, low temperature:

The entropy change associated with any condensed system undergoing a reversible isothermal process approaches zero as the temperature at which it is performed approaches 0 K.

Here a condensed system refers to liquids and solids. A classical formulation by Nernst is:

It is impossible for any process, no matter how idealized, to reduce the entropy of a system to its absolute-zero value in a finite number of operations.

There also exists a formulation of the third law which approaches the subject by postulating a specific energy behavior:

If the composite of two thermodynamic systems constitutes an isolated system, then any energy exchange in any form between those two systems is bounded.

Nernst Theorem

The third law of thermodynamics is concerned with the limiting behavior of systems as the temperature approaches zero. The bulk of the thermodynamics does not require this postulate because in thermodynamics calculations usually only entropy differences are used. Consequently, the zero point of the entropy scale is often not important. However, we discuss the third law at this point because it is it closes the postulatory basis of thermodynamics.

According to equation the temperature is defined as

Therefore, the third law states that

lim S = 0

T®0

There are several ways to state the third law of thermodynamics. It turns out that all of them are equivalent, and that one can derive one from the other. Let us start with the following form, a statement that summarizes a lot of experimental observations:

"It is impossible reduce the temperature of any systems to absolute zero in a finite number of steps."

Let's discuss this in more detail.

Assume a system to be cooled by varying a parameter X from the initial state i to the finale state f Xi to Xf. This cools the system from the temperature Ti to Tf. Using only the second law we can write for the entropy of the initial state

and for the final state

Equations can be rewritten as:

Heat capacities are positive. Thus, maximum cooling can be obtained only if the process if reversible and only without thermal contact to the environment, i.e. adiabatically. Reversibility implies:

According to the third law

$$S(0,X_i) = S(0,Xf)$$

The temperature of the final state is zero, i.e., Tf = 0. This implies:

This is impossible. We showed that absolute zero temperature can not be achieved in a finite step and, consequently, on a finite number of steps.

Equilibrium in an External Force Field

Consider a general laminar object which is free to pivot about a fixed perpendicular axis. Assuming that the object is placed in a uniform gravitational field, what is the object's equilibrium configuration in this field?

Let O represent the pivot point, and let C be the centre of mass of the object. See Fig. 90. Suppose that r represents the distance between points O and C, whereas , is the angle subtended between the line OC and the downward vertical. There are two external forces acting on the object. First, there is the downward force,Mg, due to gravity, which acts at the centre of mass. Second, there is the reaction, R, due to the pivot, which acts at the pivot point. Here, M is the mass of the object, and g is the acceleration due to gravity.

Two conditions must be satisfied in order for a given configuration of the object shown in to represent an equilibrium configuration. First, there must be zero net external force acting on the object. This implies that the reaction, R, is equal and opposite to the gravitational force, Mg. In other words, the reaction is of magnitude Mg and is directed vertically upwards. The second condition is that there must be zero net torque acting about the pivot point. Now, the reaction, R, does not generate a torque, since it acts at the pivot point. Moreover, the torque associated with the gravitational force, Mg, is simply the magnitude of this force times the length of the lever arm, d. Hence, the net torque acting on the system about the pivot point is T = M g d = M g r Sin ,.

Setting this torque to zero, we obtain sin ,, which implies that , =0o. In other words, the equilibrium configuration of a general laminar object (which is free to rotate about a fixed perpendicular axis in a uniform gravitational field) is that in which the centre of mass of the object is aligned vertically below the pivot point.

Incidentally, we can use the above result to experimentally determine the centre of mass of a given laminar object. We would need to suspend the object from two different pivot points, successively. In each equilibrium configuration, we would mark a line running vertically downward from the pivot point, using a plumb-line. The crossing point of these two lines would indicate the position of the centre of mass.

Our discussion of the equilibrium configuration of the laminar object shown in Fig. 90 is not quite complete. We have determined that the condition which must be satisfied by an equilibrium state is sin , = 0. However, there are, in fact, two physical roots of this equation. The first, , =0o, corresponds to the case where the centre of mass of the object is aligned vertically below the pivot point. The second, , =180o, corresponds to the case where the centre of mass is aligned vertically above the pivot point. Of course, the former root is far more important than the latter, since the former root corresponds to a stable equilibrium, whereas the latter corresponds to an unstable equilibrium.

We recall, from Sect., that when a system is slightly disturbed from a stable equilibrium then the forces and torques which act upon it tend to return it to this equilibrium, and vice versa for an unstable equilibrium. The easiest way to distinguish between stable and unstable equilibria, in the present case, is to evaluate the gravitational potential energy of the system. The potential energy of the object shown in, calculated using the height of the pivot as the reference height, is simply U = -Ngh = - Mgr cos ,

It can be seen that , = 0o corresponds to a minimum of this potential, whereas , = 180o corresponds to a maximum. This is in accordance with Sect. where it was demonstrated that whenever an object moves in a conservative force-field (such as a gravitational field), the stable equilibrium points correspond to minima of the potential energy associated with this field, whereas the unstable equilibrium points correspond to maxima

5

POWER TRANSMISSION AND SAFETY

You walk into your house today, flick a switch and you expect a light to come on. You press a button on the remote control and you expect the TV to come on. There are many things you depend on electricity to do each day, from running your well pump if you live in the country to washing and drying your cloths. So, how did people get by before they got electricity?

Electrical power is a little bit like the air you breathe: You don't really think about it until it is missing. Power is just "there," meeting your every need, constantly. It is only during a power failure, when you walk into a dark room and instinctively hit the useless light switch that you realize how important power is in your daily life. You use it for heating, cooling, cooking, refrigeration, light, sound, computation, entertainment... Without it, life can get somewhat cumbersome. Power travels from the power plant to your house through an amazing system called the power distribution grid.

POWER TRANSMISSION

It is the movement of energy from its place of generation to a location where it is applied to performing useful work.

Power is defined formally as units of energy per unit time. In SI units:

Since the development of technology, transmission and storage systems have been of immense interest to technologists and technology users.

Electrical power transmission has replaced mechanical power transmission in all but the very shortest distances. From the 16th century through the industrial revolution to the end of the 19th century mechanical power transmission was the norm. The oldest long-distance power transmission technology involved systems of push-rods (stängenkunst or feldstängen) connecting waterwheels to distant mine-drainage and brine-well pumps. A surviving example from 1780 exists at Bad Kösen that transmits power approximately 200 meters from a waterwheel to a salt well, and from there, an additional 150 meters to a brine evaporator. This technology survived into the 21st century in a handful of oilfields in the US, transmitting power from a central pumping engine to the numerous pump-jacks in the oil field.

Factories were fitted with overhead line shafts providing rotary power. Short line-shaft systems were described by Agricola, connecting a waterwheel to numerous ore-processing machines. While the machines described by Agricola used geared connections from the shafts to the machinery, by the 19th century, drivebelts would become the norm for linking individual machines to the line shafts. One mid 19th century factory had 1,948 feet of line shafting with 541 pulleys.

Mechanical power may be transmitted directly using a solid structure such as a driveshaft; transmission gears can adjust the amount of torque or force vs. speed in much the same way an electrical transformer adjusts voltage vs current.

Hydraulic systems use liquid under pressure to transmit power; canals and hydroelectric power generation facilities harness natural water power to lift ships or generate electricity. Pumping water or pushing mass uphill with (windmill pumps) is one possible means of energy storage. London had a hydraulic network powered by five pumping stations operated by the London Hydraulic Power Company, with a total effect of 5 MW.

Pneumatic systems use gasses under pressure to transmit power; compressed air is commonly used to operate pneumatic tools in factories and repair garages. A pneumatic wrench (for instance) is used to remove and install automotive tires far more quickly than could be done with standard manual hand tools.

A pneumatic system was proposed by proponents of Edison's direct current as the basis of the power grid. Compressed air generated at Niagara Falls would drive far away generators of DC power. The War of Currents ended with alternating current (AC) as the only means of long distance power transmission.

Electric Power Transmission

It is the bulk transfer of electrical energy, from generating power plants to substations located near to population centers. This is distinct from the local wiring between high voltage substations and customers, which is typically referred to as electricity distribution.

Transmission lines, when interconnected with each other, become high voltage transmission networks. In the US, these are typically referred to as "power grids" or just "the grid", while in the UK the network is known as the "national grid." North America has three major grids: The Western Interconnection; The Eastern Interconnection and the Electric Reliability Council of Texas (or ERCOT) grid.

Historically, transmission and distribution lines were owned by the same company, but over the last decade or so many countries have liberalized the electricity market in ways that have led to the separation of the electricity transmission business from the distribution business.

Transmission lines mostly use three-phase alternating current (AC), although single phase AC is sometimes used in railway electrification systems. High-voltage direct-current (HVDC) technology is used only for very long distances (typically greater than 400 miles, or 600 km); submarine power cables (typically longer than 30 miles, or 50 km); or for connecting two AC networks that are not synchronized.

Electricity is transmitted at high voltages (110 kV or above) to reduce the energy lost in long distance transmission. Power is usually transmitted through overhead power lines. Underground power transmission has a significantly higher cost and greater operational limitations but is sometimes used in urban areas or sensitive locations.

A key limitation in the distribution of electricity is that, with minor exceptions, electrical energy cannot be stored, and therefore must be generated as needed. A sophisticated system of control is therefore required to ensure electric generation very closely matches the demand. If supply and demand are not in balance, generation plants and transmission equipment can shut down which, in the worst cases, can lead to a major regional blackout, such as occurred in California and the US Northwest in 1996 and in the US Northeast in 1965, 1977 and 2003.

To reduce the risk of such failures, electric transmission networks are interconnected into regional, national or continental wide networks thereby providing multiple redundant alternate routes for power to flow should (weather or equipment) failures occur. Much analysis is done by transmission companies to determine the maximum reliable capacity of each line which is mostly less than its physical or thermal limit, to ensure spare capacity is available should there be any such failure in another part of the network.

Electrical Safety

We rely on electricity to light our homes, keep our food hot or cold, run our appliances, charge our phones, and so much more. Electricity, like many things we use so often, is sometimes taken for granted and overlooked as something that could potentially be very hazardous. It's important to know and practice proper safety when it comes to something as deadly as electricity.

Make sure the power is off. Shut off the breakers in the breaker box. To make extra sure, use a tester before touching any wiring. Never touch bare wiring with any part of your body. Use non-conductive shoes and gloves when working with electricity. When working with tools, be sure that the tools are properly insulated and do not have metal handles. Also, be sure to never work on electrical repairs that are in a wet area.

Even when wires are not exposed, electricity can still be very dangerous. Keep all electrical devices away from parts of the home that use water on a regular basis such as bath tubs, showers and sinks. When using things like hair dryers or curlers in the bathroom, make sure the area around you is dry, and be sure to unplug the device when you are done. Electrical devices can cause sparks, especially if they're knocked around or become broken. These sparks are hot enough to start fires on carpet, curtains, clothes, bed sheets and the like. Be sure all running electrical devices are in a safe and stable place.

One of the most threatening electrical hazards outside of the home is a downed power line. You are unable to turn off the power to these lines and can be killed if you come in contact with them. If you find yourself close to a downed power line, leave the area immediately and call 9-1-1. If a lot of people are around the power line, try to help by letting everyone know they need to stay away.

Innocent Children

Many times, children are not aware of the dangers of electricity. They can sometimes find it intriguing, sparking their curiosity. To prevent children from playing with power outlets, find and purchase outlet covers for any outlets not in use. Also, be sure to teach your children not to play with cords that are occupying outlets. Older kids will also be curious with electrical devices and will sometimes want to open them and take them apart. Teach your children not to do this as even unplugged devices can sometimes hold a current strong enough to cause harm.

Define Overhead power line

An overhead power line is an electric power transmission line suspended by towers or utility poles. Since most of the insulation is provided by air, overhead power lines are generally the lowest-cost method of transmission for large quantities of electric energy. Towers for support of the lines are made of wood (as-grown or laminated), steel (either lattice structures or tubular poles), concrete, aluminium, and occasionally reinforced plastics.

The bare wire conductors on the line are generally made of aluminium (either plain or reinforced with steel or sometimes composite materials), though some copper wires are used in medium-voltage distribution and low-voltage connections to customer premises. A major goal of overhead power line design is to maintain adequate clearance between energized conductors and the ground so as to prevent dangerous contact with the line.

The invention of the strain insulator was a critical factor in allowing higher voltages to be used. At the end of the 19th century, the limited electrical strength of telegraph-style pin insulators limited the voltage to no more than 69,000 volts. Today overhead lines are routinely operated at voltages exceeding 765,000 volts between conductors, with even higher voltages possible in some cases.

Overhead power transmission lines are classified in the electrical power industry by the range of voltages:

- Low voltage - less than 1000 volts, used for connection between a residential or small commercial customer and the utility.
- Medium Voltage (Distribution) - between 1000 volts (1 kV) and to about 33 kV, used for distribution in urban and rural areas.
- High Voltage (subtransmission less than 100 kV; subtransmission or transmission at voltage such as 115 kV and 138 kV), used for sub-transmission and transmission of bulk quantities of electric power and connection to very large consumers.
- Extra High Voltage (transmission) - over 230 kV, up to about 800 kV, used for long distance, very high power transmission.
- Ultra High Voltage - higher than 800 kV.

Structures for overhead lines take a variety of shapes depending on the type of line. Structures may be as simple as wood poles directly set in the earth, carrying one or more cross-arm beams to support conductors, or "armless" construction with conductors supported on insulators attached to the side of the pole. Tubular steel poles are typically used in urban areas. High-voltage lines are often carried on lattice-type steel towers or pylons. For remote areas, aluminium towers may be placed by helicopters. Concrete poles have also been used. Poles made of reinforced plastics are also available, but their high cost restricts application.

Each structure must be designed for the loads imposed on it by the conductors. A large transmission line project may have several types of towers, with "tangent" ("suspension" or "line" towers, UK) towers intended for most positions and more heavily constructed towers used for turning the line through an angle, dead-ending (terminating) a line, or for important river or road crossings. Depending on the design criteria for a particular line, semi-flexible type structures may rely on the weight of the conductors to be balanced on both sides of each tower. More rigid structures may be intended to remain standing even if one or more conductors is broken. Such structures may be installed at intervals in power lines to limit the scale of cascading tower failures.

Foundations for tower structures may be large and costly, particularly if the ground conditions are poor, such as in wetlands. Each structure may be stabilized considerably by the use of guy wires to counteract some of the forces applied by the conductors.

Power lines and supporting structures can be a form of visual pollution. In some cases the lines are buried to avoid this, but this "undergrounding" is more expensive and therefore not common.

For a single wood utility pole structure, a pole is placed in the ground, then three crossarms extend from this, either staggered or all to one side. The insulators are attached to the crossarms. 1For an "H"-type wood pole structure, two poles are placed in the ground, then a crossbar is placed on top of these, extending to both sides. The insulators are attached at the ends and in the middle. Lattice tower structures have two common forms. One has a pyramidal base, then a vertical section, where three crossarms extend out, typically staggered. The strain insulators are attached to the crossarms. Another has a pyramidal base, which extends to four support points. On top of this a horizontal truss-like structure is placed. The insulators are attached to this.

INSULATORS

Insulators must support the conductors and withstand both the normal operating voltage and surges due to switching and lightning. Insulators are broadly classified as either pin-type, which support the conductor above the structure, or suspension type, where the conductor hangs below the structure. Up to about 33 kV (69 kV in North America) both types are commonly used.

At higher voltages only suspension-type insulators are common for overhead conductors. Insulators are usually made of wet-process porcelain or toughened glass, with increasing use of glass-reinforced polymer insulators. However, with rising voltage levels and changing climatic conditions, polymer insulators (silicone rubber based) are seeing increasing usage. China has already developed polymer insulators having a highest system voltage of 1100kV and India is currently developing a 1200kV (highest system voltage) line which will initially be charged with 400kV to be upgraded to a 1200kV line.

Suspension insulators are made of multiple units, with the number of unit insulator disks increasing at higher voltages. The number of disks is chosen based on line voltage, lightning withstand requirement, altitude, and environmental factors such as fog, pollution, or salt spray. Longer insulators, with longer creepage distance for leakage current, are required in these cases. Strain insulators must be strong enough mechanically to support the full weight of the span of conductor, as well as loads due to ice accumulation, and wind.

Porcelain insulators may have a semi-conductive glaze finish, so that a small current (a few milliamperes) passes through the insulator. This warms the surface slightly and reduces the effect of fog and dirt accumulation. The semiconducting glaze also ensures a more even distribution of voltage along the length of the chain of insulator units.

Polymer insulators by nature have hydrophobic characteristics providing for improved wet performance. Also, studies have shown that the specific creepage distance required in polymer insulators is much lower than that required in porcelain or glass. Additionally, the mass of polymer insulators (espicially in higher voltages) is approximately 50% to 30% less than that of a comparative porcelain or glass string. Better pollution and wet performance is leading to the increased use of such insulators.

Insulators for very high voltages, exceeding 200 kV, may have grading rings installed at their terminals. This improves the electric field distribution around the insulator and makes it more resistant to flash-over during voltage surges.

Conductors

Aluminium conductors reinforced with steel (known as ACSR) are primarily used for medium and high voltage lines and may also be used for overhead services to individual customers. Aluminium conductors are used as it has the advantage of lower resistivity/weight than copper, as well as being cheaper. Some copper cable is still used, especially at lower voltages and for grounding.

While larger conductors may lose less energy due to lower electrical resistance, they are more costly than smaller conductors. An optimization rule called Kelvin's Law states that the optimum size of conductor for a line is found when the cost of the energy wasted in the conductor is equal to the annual interest paid on that portion of the line construction cost due to the size of the conductors. The optimization problem is made more complex due to additional factors such as varying annual load, varying cost of installation, and by the fact that only definite discrete sizes of cable are commonly made.

Since a conductor is a flexible object with uniform weight per unit length, the geometric shape of a conductor strung on towers approximates that of a catenary. The sag of the conductor (vertical distance between the highest and lowest point of the curve) varies depending on the temperature.

A minimum overhead clearance must be maintained for safety. Since the temperature of the conductor increases with increasing heat produced by the current through it, it is sometimes possible to increase the power handling capacity (uprate) by changing the conductors for a type with a lower coefficient of thermal expansion or a higher allowable operating temperature.

Bundle Conductors

Bundle conductors are used to reduce corona losses and audible noise. Bundle conductors consist of several conductor cables connected by non-conducting spacers. For 220 kV lines, two-conductor bundles are usually used, for 380 kV lines usually three or even four. American Electric Power is building 765 kV lines using six conductors per phase in a bundle. Spacers must resist the forces due to wind, and magnetic forces during a short-circuit.

Bundle conductors are used to increase the amount of current that may be carried in a line. Due to the skin effect, ampacity of conductors is not proportional to cross section, for the larger sizes. Therefore, bundle conductors may carry more current for a given weight.

A bundle conductor results in lower reactance, compared to a single conductor. It reduces corona discharge loss at EHV (extra high voltage) and interference with communication systems. It also reduces voltage gradient in that range of voltage. As a disadvantage, the bundle conductors have higher wind loading.

Transmission Line

A single-circuit transmission line carries conductors for only one circuit. For a three-phase system, this implies that each tower supports three conductors.

A double-circuit transmission line has two circuits. For three-phase systems, each tower supports and insulates six conductors. Single phase AC-powerlines as used for traction current have four conductors for two circuits. Usually both circuits operate at the same voltage. In HVDC systems typically two conductors are carried per line, but rarely only one pole of the system is carried on a set of towers.

In some countries like Germany most powerlines with voltages above 100 kV are implemented as double, quadruple or in rare cases even hexuple powerline as rights of way are rare. Sometimes all conductors are installed with the erection of the pylons; often some circuits are installed later.

A disadvantage of double circuit transmission lines is that maintenance works can be more difficult, as either work in close proximity of high voltage or switch-off of 2 circuits is required. In case of failure, both systems can be affected.The largest double-circuit transmission line is the Kita-Iwaki Powerline.

Ground wires

Overhead power lines are often equipped with a ground conductor (shield wire or overhead earth wire). A ground conductor is a conductor that is usually grounded (earthed) at the top of the supporting structure to minimise the likelihood of direct lightning strikes to the phase conductors. The ground wire is also a parallel path with the earth for fault currents in earthed neutral circuits. Very high-voltage transmission lines may have two ground conductors. These are either at the outermost ends of the highest cross beam, at two V-shaped mast points, or at a separate cross arm. Older lines may use surge arrestors every few spans in place of a shield wire; this configuration is typically found in the more rural areas of the United States. By protecting the line from lightning, the design of apparatus in substations is simplified due to lower stress on insulation. Shield wires on transmission lines may include optical fibers (OPGW), used for communication and control of the power system.

Medium-voltage distribution lines may have the grounded conductor strung below the phase conductors to provide some measure of protection against tall vehicles or equipment touching the energized line, as well as to provide a neutral line in Wye wired systems.

USE OF INSULATED CONDUCTORS

While overhead lines are usually bare conductors, rarely overhead insulated cables are used, usually for short distances (less than a kilometer). Insulated cables can be directly fastened to structures without insulating supports. An overhead line with bare conductors insulated by air is typically less costly than a cable with insulated conductors.

A more common approach is "covered" line wire. It is treated as bare cable, but often is safer for wildlife, as the insulation on the cables increases the likelihood of a large wing-span raptor to survive a brush with the lines, and reduces the overall danger of the lines slightly. These types of lines are often seen in the eastern United States and in heavily wooded areas, where tree-line contact is likely.

The only pitfall is cost, as insulated wire is often costlier than its bare counterpart. Many utility companies implement covered line wire as jumper material where the wires are often closer to each other on the pole, such as an underground riser/Pothead, and on reclosers, cutouts and the like.

Low voltage

Low voltage overhead lines may use either bare conductors carried on glass or ceramic insulators or an aerial bundled cable system. The number of conductors may be anywhere between four (three phase plus a combined earth/neutral conductor - a TN-C earthing system) up to as many as six (three phase conductors, separate neutral and earth plus street lighting supplied by a common switch).

Train power

Overhead lines or overhead wires are used to transmit electrical energy to trams, trolleybuses or trains. Overhead line is designed on the principle of one or more overhead wires situated over rail tracks. Feeder stations at regular intervals along the overhead line supply power from the high voltage grid. For some cases low-frequency AC is used, and distributed by a special traction current network.

Further applications

Overhead lines are also occasionally used to supply transmitting antennas, especially for efficient transmission of long, medium and short waves. For this purpose a staggered array line is often used. Along a staggered array line the conductor cables for the supply of the earth net of the transmitting antenna are attached on the exterior of a ring, while the conductor inside the ring, is fastened to insulators leading to the high voltage standing feeder of the antenna.

Usage of area under overhead power lines

Use of the area below an overhead line is restricted because objects must not come too close to the energized conductors. Overhead lines and structures may shed ice, creating a hazard. Radio reception can be impaired under a power line, due both to shielding of a receiver antenna by the overhead conductors, and by partial discharge at insulators and sharp points of the conductors which creates radio noise.

In the area surrounding overhead lines it is dangerous to risk interference; e.g. flying kites or balloons, using ladders or operating machinery.

Overhead distribution and transmission lines near airfields are often marked on maps, and the lines themselves marked with conspicuous plastic reflectors, to warn pilots of the presence of conductors.

Construction of overhead power lines, especially in wilderness areas, may have significant environmental effects. Environmental studies for such projects may consider the effect of brush clearing, changed migration routes for migratory animals, possible access by predators and humans along transmission corridors, disturbances of fish habitat at stream crossings, and other effects.

HISTORY

The first transmission of electrical impulses over an extended distance was demonstrated on July 14, 1729 by the physicist Stephen Gray, in order to show that one can transfer electricity by that method. The demonstration used damp hemp cords suspended by silk threads (the low resistance of metallic conductors not being appreciated at the time).

However the first practical use of overhead lines was in the context of telegraphy. By 1837 experimental commercial telegraph systems ran as far as 13 miles (20 km). Electric power transmission was accomplished in 1882 with the first high voltage transmission between Munich and Miesbach. 1891 saw the construction of the first three-phase alternating current overhead line on the occasion of the International Electricity Exhibition in Frankfurt, between Lauffen and Frankfurt.

In 1912 the first 110 kV-overhead power line entered service followed by the first 220 kV-overhead power line in 1923. In the 1920s RWE AG built the first overhead line for this voltage and in 1926 built a Rhine crossing with the pylons of Voerde, two masts 138 meters high.

In Germany in 1957 the first 380 kV overhead power line was commissioned (between the transformer station and Rommerskirchen). In the same year the overhead line traversing of the Strait of Messina went into service in Italy, whose pylons served the Elbe crossing 1. This was used as the model for the building of the Elbe crossing 2 in the second half of the 1970s which saw the construction of the highest overhead line pylons of the world.

Starting from 1967 in Russia, and also in the USA and Canada, overhead lines for voltage of 765 kV were built. In 1982 overhead power lines were built in Russia between Elektrostal and the power station at Ekibastusz, this was a three-phase alternating current line at 1150 kV (Powerline Ekibastuz-Kokshetau). In 1999, in Japan the first powerline designed for 1000 kV with 2 circuits were built, the Kita-Iwaki Powerline. In 2003 the building of the highest overhead line commenced in China, the Yangtze River Crossing.

Personal Safety during Electrical Storms

Modern science is still studying lightning in order to better understand it. Much more still remains to be learned about it. It was only a few decades ago that a statistical analysis was done regarding the time of day when lightning strokes occur. Scientists were surprised that a tremendous majority of lightning strokes occur between around noon and 4 p.m. in nearly all locations. It is thought that this is due to there being far more energy available in the atmosphere during those hours due to bright sunshine providing that energy. It seems likely that heating of the ground with such solar heating causes convection of the atmosphere above it so that air is rapidly rising. Such rapid vertical air motion can aid in segregating positive and negative charges in clouds at the upper and lower altitudes. This seems like a logical method where the amazingly high voltages and currents that cause lightning strokes could develop.

If anywhere near the current of an actual lightning bolt passed through any living body, the moisture inside all of the cells would instantly boil and the person or animal would explode. When trees get hit, this generally happens, and the exploding water vapor inside the many cells of the tree causes even a strong tree to suddenly split down the middle (with a very loud sound of the water boiling and essentially exploding).

The lightning itself doesn't split the tree, the instantly boiling water in the cells of the tree does. Clearly, a human, filled with electrically conducting blood and salt water, couldn't possibly survive anything like this. It is possible to greatly minimize the chance of danger in lightning storms. Principles of Physics are used here to understand the situation better that most media present. This analysis then suggests certain logical precautions.Every second, about 100 lightning strokes occur somewhere in the Earth's atmosphere. Every year, between 100 and 200 people in the U.S. die as result of lightning strikes, with quite a few more hospitalized.

We actually are amazingly sensitive to electricity. Our brain and nervous system can only operate with some natural body electricity. It has been very well established that even a few milli-amperes (.001 ampere) passing through the chest cavity of any person or animal will stop the heart from getting its triggering signals, and the person can die. This fatal situation involves a small fraction of an ampere, far less than the 50,000 amperes and up to 1,000,000,000 volts of a lightning bolt. Even the extremely high resulting temperatures, as high as 54,000°F, would cause all the water in a person to boil and explode, which is what happens when a tree gets hit by lightning.

So, how does someone "get hit by lightning" and live? Well, they actually didn't get hit! It turns out that Physics and electromagnetism has the explanation for us. In a nutshell, often, the person became part of an "air-core transformer." Whenever electricity is passing through a conductor, (whether a water-filled tree, or any metal conductor, or even moisture-filled air) it creates a temporary magnetic field in the area surrounding the conductor. This is called electromagnetism.

If a second parallel conductor is nearby, that changing magnetic field causes an electric current to be "induced" in the second conductor, called an EMF. This is also electromagnetism. This is actually the principle of all electrical transformers and many other modern electrical devices. This particular arrangement is called an air-core transformer. In the situation we are considering here, the lightning bolt is passing through a low resistance path between the ground and the cloud (the first conductor, such as including a water-filled tree). A standing person, who is also mostly (salt) water, is a pretty good conductor of electricity, represents the second conductor. IF the person's body is relatively parallel to the first conductor, an induced current is created inside the body of the person, by that transformer effect. This current can be quite substantial, and can still be plenty high enough to cause death.

There have been many cases where many people in a crowd have all "been hit by lightning" at the same moment. A well-documented case was from a 1998 soccer match that was being televised. A lightning stroke hit the field, not actually where anyone was standing, but many of the players immediately collapsed to the ground. They were all victims of this air-core transformer phenomenon. The severity of injuries in such situations depends on several things. The most important is the distance between where the lightning stroke hit and the person, because the effectiveness of the air-core transformer phenomenon depends tremendously on the spacing between the two conductors. The second most important is the orientation of the person. If a player had been lying down on the ground that day, he likely would have not had any serious injuries at all! Of course, a soccer player who is laying down gets replaced rather quickly, so they were all standing up. There are many other possible variables, such as whether a player was is good electrical contact with wet ground through sweaty (electrically conductive) wet shoes. Fortunately, in the case of that soccer match, none of the players died, although some had severe injuries due to electrical burns. It seems strange that they could have electrical burns without actually having been hit by the lightning, but that is an indication of how effective an air-core transformer can be. It was not reported, but it is certain that every person in the stands watching that game would also have felt an intense electrical effect due to an air-core transformer effect. They were each farther away from the lightning stroke, they were not sweat covered, and they were probably fairly dry, so it is far less likely that any would have sustained any significant injuries. They were probably sitting down, where the "effective length of the second conductor" (their bodies) was shorter than the standing players on the field. It is extremely likely that they each person in the stadium felt an intense tingling throughout their bodies. It could easily have been that the standing peanut vendors might have sustained more serious injuries.

If the person has a good electrical path to ground, such as standing in water, that current can pass quickly out of the body and create serious burns. This explanation also clarifies why people holding a golf club or metal umbrella up above their head increases the chance of them "being hit." What is actually happening is that their actions just make the "second conductor" referred to above to be extended in both height and conductivity by holding such things. More of the created magnetic field lines intercept the body and umbrella, so the induced current and voltage are therefore much greater and more damaging in the body of the person. This all means that, even though the person was not actually hit by the lightning bolt, scorched shoes, skin burn marks, and other physical damage are very real. Since a metal golf club can intercept a lot of the electromagnetic effects and then conduct the resulting electricity extremely well, a golfer can receive very serious burns on the hands as a result.

Of course, if that person who is holding a metal golf club above his head is standing alone at the top of a hill, where he is the highest object around, there IS the possibility that he might actually be hit by a lightning stroke, but that would instantly stop his heart, boil all the water in his body, and kill him instantly. Actually, under those really unfortunate circumstances, there could conceivably be very little left to identify as a human.

Thunder

Whenever a lightning stroke occurs, whether from cloud to cloud or from cloud to ground, the flash of light and this shock wave in the air are created at the exact same time. If that occurred two miles from you, the light will get to you in 2/186000 seconds, since the speed of light is 186,000 miles per second. That means that the light gets to you in far less than one thousandth of a second, effectively instantly. The shock wave in the air, which has the effect of a sound wave, takes around 10 seconds to travel the 10,560 feet of the two miles.

Therefore, we see a bright lightning stroke, and then, ten seconds later, hear the rumble of thunder, which is the effect of the shock wave. That's why people sometimes count seconds after a lightning stroke to estimate the distance away. For each five seconds until the thunder, it means the lightning was another mile further away.

Sometimes, we see something called sheet lightning. Usually, this is seen at night when there is no storm around. No thunder is usually heard with sheet lightning. That doesn't mean that the thunder does not exist. This phenomenon is actually normal lightning, associated with a storm that is more than about 30 miles away. We see the effect of the light of many lightning strokes, but the sound of the thunder does not carry for the 30 or 50 or 100 miles, so we do not notice any thunder.

Thunder generally has a very low-pitched sound, because the shock wave produced by the rapidly expanding air created one very large shock wave. The physically large dimensions of that wave cause the sound to have such a rumbling low pitch. There are some variations on the sound of thunder. If a lightning stroke was from side to side (from where you see it), it might all have been at about the same distance from you, so the sound of the shock waves created all along its length would all get to you at about the same time. The sound would be a very brief but loud boom! If a lightning stroke of a mile long occurred sort of toward and away from you (in the sky), the sound from the shock waves along different parts of it (all created at the same time) would get to you spread out over about five seconds, making it sound like an extended rumbling sound, that's usually not as loud as the boom kind.

When lightning hits a tree, there is often a loud crack sound at the very start of the thunder, but usually only if the tree was nearby. This higher pitched sound is not actually from the lightning stroke itself, but is due to the sound of parts of the tree exploding as the water within it suddenly boiled.

Coupling

A coupling is a device used to connect two shafts together at their ends for the purpose of transmitting power. Couplings do not normally allow disconnection of shafts during operation, however there are torque limiting couplings which can slip or disconnect when some torque limit is exceeded.

The primary purpose of couplings is to join two pieces of rotating equipment while permitting some degree of misalignment or end movement or both. By careful selection, installation and maintenance of couplings, substantial savings can be made in reduced maintenance costs and downtime.

Shaft couplings are used in machinery for several purposes, the most common of which are the following.

- To provide for the connection of shafts of units that are manufactured separately such as a motor and generator and to provide for disconnection for repairs or alternations.
- To provide for misalignment of the shafts or to introduce mechanical flexibility.
- To reduce the transmission of shock loads from one shaft to another.
- To introduce protection against overloads.
- To alter the vibration characteristics of rotating units.

Types of shaft couplings

Rigid coupling

Rigid couplings are used when precise shaft alignment is required; shaft misalignment will affect the coupling's performance as well as its life. Examples:

- Sleeve or muff coupling
- Clamp or split-muff or compression coupling
- Flange coupling

Flexible coupling

Flexible couplings are designed to transmit torque while permitting some radial, axial, and angular misalignment. Flexible couplings can accommodate angular misalignment up to a few degrees and some parallel misalignment. Examples:

- Universal joint or Cardan joint
- Oldham coupling
- Constant-velocity joint
    - Rzeppa joint
    - Double cardan joint
    - Thompson coupling
- Bellows coupling - low backlash.
- Elastomeric coupling
    - Bushed pin coupling
    - Donut coupling
    - Spider or jaw coupling (or lovejoy coupling)

- Resilient coupling
- Disc coupling, by definition, transmits torque from a driving to a driven bolt tangentially on a common bolt circle. Torque is transmitted between the bolts through a series of thin, stainless steel discs assembled in a pack. Misalignment is accomplished by deforming of the material between the bolts.
    - Rag joint
    - Giubo
- Diaphragm coupling transmits torque from the outside diameter of a flexible plate to the inside diameter, across the spool or spacer piece, and then from inside to outside diameter. The deforming of a plate or series of plates from I.D. to O.D accomplishes the misalignment.
- Gear coupling
- Hirth joint
- Roller chain and sprocket coupling
- Fluid coupling

Requirements of good shaft alignment / good coupling setup

- It should be easy to connect or disconnect the coupling.
- It should transmit the full power from one shaft to other without losses.
- It does allow some misalignment between the two adjacent shaft rotation axis.
- It is the goal to minimise the remaining misalignment in running operation to maximise power transmission and to maximise machine runtime (coupling and bearing and sealings lifetime).
- It should have no projecting parts.
- It is recommended to use manufacturer's alignment target values to set up the machine train to a defined non-zero alignment, due to the fact that later when the machine is at operation temperature the alignment condition is perfect

Tools to measure shaft axis alignment condition

- It is possible to measure the alignment with dial gauges or feeler gauges using various mechanical setups.

- It is recommended to take care of bracket sag, parallaxe error while reading the values.
- It is very convenient to use laser shaft alignment technique to perform the alignment task within highest accuracy.
- It is required to align the machine better, the laser shaft alignment tool can help to show the required moves at the feet positions.

Coupling maintenance and failure

Coupling maintenance is generally a simple matter, requiring a regularly scheduled inspection of each coupling. It consists of:

- Performing visual inspections, checking for signs of wear or fatigue, and cleaning couplings regularly.
- Checking and changing lubricant regularly if the coupling is lubricated. This maintenance is required annually for most couplings and more frequently for couplings in adverse environments or in demanding operating conditions.
- Documenting the maintenance performed on each coupling, along with the date.

Even with proper maintenance, however, couplings can fail. Underlying reasons for failure, other than maintenance, include:

- Improper installation
- Poor coupling selection
- Operation beyond design capabilities.

The only way to improve coupling life is to understand what caused the failure and to correct it prior to installing a new coupling. Some external signs that indicate potential coupling failure include:

- Abnormal noise, such as screeching, squealing or chattering
- Excessive vibration or wobble
- Failed seals indicated by lubricant leakage or contamination.

CHECKING THE COUPLING BALANCE

Couplings are normally balanced at the factory prior to being shipped, but they occasionally go out of balance in operation. Balancing can be difficult and expensive, and is normally done only when operating tolerances are such that the effort and the expense are justified. The amount of coupling unbalance that can be tolerated by any system is dictated by the characteristics of the specific connected machines and can be determined by detailed analysis or experience.

Welding

Welding is a fabrication or sculptural process that joins materials, usually metals or thermoplastics, by causing coalescence. This is often done by melting the workpieces and adding a filler material to form a pool of molten material (the weld pool) that cools to become a strong joint, with pressure sometimes used in conjunction with heat, or by itself, to produce the weld. This is in contrast with soldering and brazing, which involve melting a lower-melting-point material between the workpieces to form a bond between them, without melting the workpieces.

Many different energy sources can be used for welding, including a gas flame, an electric arc, a laser, an electron beam, friction, and ultrasound. While often an industrial process, welding can be done in many different environments, including open air, under water and in outer space. Regardless of location, welding remains dangerous, and precautions are taken to avoid burns, electric shock, eye damage, poisonous fumes, and overexposure to ultraviolet light.

Arc welding

Arc welding is a type of welding that uses a welding power supply to create an electric arc between an electrode and the base material to melt the metals at the welding point. They can use either direct (DC) or alternating (AC) current, and consumable or non-consumable electrodes. The welding region is sometimes protected by some type of inert or semi-inert gas, known as a shielding gas, and/or an evaporating filler material. The process of arc welding is widely used because of its low capital and running costs. Getting the arc started is called striking the arc. An arc may be struck by either lightly tapping the electrode against the metal or scratching the electrode against the metal at high speed.

Competing welding processes such as resistance welding and oxyfuel welding were developed during this time as well; but both, especially the latter, faced stiff competition from arc welding especially after metal coverings (known as flux) for the electrode, to stabilize the arc and shield the base material from impurities, continued to be developed.

Porosity and brittleness were the primary problems and the solutions that developed included the use of hydrogen, argon, and helium as welding atmospheres. During the following decade, further advances allowed for the welding of reactive metals such as aluminum and magnesium. This, in conjunction with developments in automatic welding, alternating current, and fluxes fed a major expansion of arc welding during the 1930s and then during World War II.

During the middle of the century, many new welding methods were invented. Submerged arc welding was invented in 1930 and continues to be popular today. In 1932 a Russian, Konstantin Khrenov successfully implemented the first underwater electric arc welding. Gas tungsten arc welding, after decades of development, was finally perfected in 1941 and gas metal arc welding followed in 1948, allowing for fast welding of non-ferrous materials but requiring expensive shielding gases.

Using a consumable electrode and a carbon dioxide atmosphere as a shielding gas, it quickly became the most popular metal arc welding process. In 1957, the flux-cored arc welding process debuted in which the self-shielded wire electrode could be used with automatic equipment, resulting in greatly increased welding speeds. In that same year, plasma arc welding was invented. Electroslag welding was released in 1958 and was followed by its cousin, electrogas welding, in 1961.

Power supplies

To supply the electrical energy necessary for arc welding processes, a number of different power supplies can be used. The most common classification is constant current power supplies and constant voltage power supplies. In arc welding, the voltage is directly related to the length of the arc, and the current is related to the amount of heat input. Typical currents are 50 to 500 amps, depending on the size of weld required; 100 amps is typical for manual welders. Voltage output is typically 20 to 50 volts during welding, though some power supplies also include a small high voltage source to aid in initially striking the arc.

Constant current power supplies are most often used for manual welding processes such as gas tungsten arc welding and shielded metal arc welding, because they maintain a relatively constant current even as the voltage varies. This is important because in manual welding, it can be difficult to hold the electrode perfectly steady, and as a result, the arc length and thus voltage tend to fluctuate.

Constant voltage power supplies hold the voltage constant and vary the current, and as a result, are most often used for automated welding processes such as gas metal arc welding, flux cored arc welding, and submerged arc welding. In these processes, arc length is kept constant, since any fluctuation in the distance between the wire and the base material is quickly rectified by a large change in current. For example, if the wire and the base material get too close, the current will rapidly increase, which in turn causes the heat to increase and the tip of the wire to melt, returning it to its original separation distance.

The direction of current used in arc welding also plays an important role in welding. Consumable electrode processes such as shielded metal arc welding and gas metal arc welding generally use direct current, but the electrode can be charged either positively or negatively.

In welding, the positively charged anode will have a greater heat concentration and, as a result, changing the polarity of the electrode has an impact on weld properties. If the electrode is positively charged, it will melt more quickly, increasing weld penetration and welding speed. Alternatively, a negatively charged electrode results in more shallow welds.

Non-consumable electrode processes, such as gas tungsten arc welding, can use either type of direct current (DC), as well as alternating current (AC). With direct current however, because the electrode only creates the arc and does not provide filler material, a positively charged electrode causes shallow welds, while a negatively charged electrode makes deeper welds.

Alternating current rapidly moves between these two, resulting in medium-penetration welds. One disadvantage of AC, the fact that the arc must be re-ignited after every zero crossing, has been addressed with the invention of special power units that produce a square wave pattern instead of the normal sine wave, eliminating low-voltage time after the zero crossings and minimizing the effects of the problem.

CONSUMABLE ELECTRODE METHODS

One of the most common types of arc welding is shielded metal arc welding (SMAW), which is also known as manual metal arc welding (MMA) or stick welding. An electric current is used to strike an arc between the base material and a consumable electrode rod or 'stick'. The electrode rod is made of a material that is compatible with the base material being welded and is covered with a flux that protects the weld area from oxidation and contamination by producing CO2 gas during the welding process. The electrode core itself acts as filler material, making a separate filler unnecessary. The process is very versatile, requiring little operator training and inexpensive equipment.

However, weld times are rather slow, since the consumable electrodes must be frequently replaced and because slag, the residue from the flux, must be chipped away after welding. Furthermore, the process is generally limited to welding ferrous materials, though specialty electrodes have made possible the welding of cast iron, nickel, aluminium, copper and other metals. The versatility of the method makes it popular in a number of applications including repair work and construction.

Gas metal arc welding (GMAW), commonly called MIG (Metal Inert Gas), is a semi-automatic or automatic welding process with a continuously fed consumable wire acting as both electrode and filler metal, along with an inert or semi-inert shielding gas flowed around the wire to prevent the weld site from contamination.

Constant voltage, direct current power source is most commonly used with GMAW, but constant current alternating current are used as well. With continuously fed filler electrodes, GMAW offers relatively high welding speeds, however the more complicated equipment reduces convenience and versatility in comparison to the SMAW process.

Originally developed for welding aluminium and other non-ferrous materials in the 1940s, GMAW was soon economically applied to steels. Today, GMAW is commonly used in industries such as the automobile industry for its quality, versatility and speed. Because of the need to maintain a stable shroud of shielding gas around the weld site, it can be problematic to use the GMAW process in areas of high air movement such as outdoors.

Flux-cored arc welding (FCAW) is a variation of the GMAW technique. FCAW wire is actually a fine metal tube filled with powdered flux materials. Flux cored wire generates an effective gas shield precisely at the weld site, permitting application involving more windy conditions or contaminated materials, however the flux cored wire leaves a slag residue and is more expensive than solid wire.

Submerged arc welding (SAW) is a high-productivity automatic welding method in which the arc is struck beneath a covering layer of flux. This increases arc quality, since contaminants in the atmosphere are blocked by the flux. The slag that forms on the weld generally comes off by itself and, combined with the use of a continuous wire feed, the weld deposition rate is high.

Working conditions are much improved over other arc welding processes since the flux hides the arc and no smoke is produced. The process is commonly used in industry, especially for large products. As the arc is not visible, it requires full automatization. In-position welding is not possible with SAW.

Non-consumable electrode methods

Gas tungsten arc welding (GTAW), or tungsten inert gas (TIG) welding, is a manual welding process that uses a non-consumable electrode made of tungsten, an inert or semi-inert gas mixture, and a separate filler material. Especially useful for welding thin materials, this method is characterized by a stable arc and high quality welds, but it requires significant operator skill and can only be accomplished at relatively low speeds.

It can be used on nearly all weldable metals, though it is most often applied to stainless steel and light metals. It is often used when quality welds are extremely important, such as in bicycle, aircraft and naval applications. A related process, plasma arc welding, also uses a tungsten electrode but uses plasma gas to make the arc. The arc is more concentrated than the GTAW arc, making transverse control more critical and thus generally restricting the technique to a mechanized process.

Because of its stable current, the method can be used on a wider range of material thicknesses than can the GTAW process and is much faster. It can be applied to all of the same materials as GTAW except magnesium; automated welding of stainless steel is one important application of the process. A variation of the process is plasma cutting, an efficient steel cutting process.

Other arc welding processes include atomic hydrogen welding, carbon arc welding, electroslag welding, electrogas welding, and stud arc welding.

Corrosion issues

Some materials, notably high-strength steels, aluminium, and titanium alloys, are susceptible to hydrogen embrittlement. If the electrodes used for welding contain traces of moisture, the water decomposes in the heat of the arc and the liberated hydrogen enters the lattice of the material, causing its brittleness.

Electrodes for such materials, with special low-hydrogen coating, are delivered in sealed moisture-proof packaging. New electrodes can be used straight from the can, but when moisture absorption may be suspected, they have to be dried by baking (usually at 800 to 1000 °F (425 to 550 °C)) in a drying oven. Flux used has to be kept dry as well.

Some austenitic stainless steels and nickel-based alloys are prone to intergranular corrosion. When subjected to temperatures around 700 °C (1,300 °F) for too long time, chromium reacts with carbon in the material, forming chromium carbide and depleting the crystal edges of chromium, impairing their corrosion resistance in a process called sensitization. Such sensitized steel undergoes corrosion in the areas near the welds where the temperature-time was favorable for forming the carbide. This kind of corrosion is often termed weld decay.

Knifeline attack (KLA) is another kind of corrosion affecting welds, impacting steels stabilized by niobium. Niobium and niobium carbide dissolves in steel at very high temperatures. At some cooling regimes, niobium carbide does not precipitate, and the steel then behaves like unstabilized steel, forming chromium carbide instead.

This affects only a thin zone several millimeters wide in the very vicinity of the weld, making it difficult to spot and increasing the corrosion speed. Structures made of such steels have to be heated in a whole to about 1,950 °F (1,070 °C), when the chromium carbide dissolves and niobium carbide forms. The cooling rate after this treatment is not important. Filler metal (electrode material) improperly chosen for the environmental conditions can make them corrosion-sensitive as well. There are also issues of galvanic corrosion if the electrode composition is sufficiently dissimilar to the materials welded, or the materials are dissimilar themselves. Even between different grades of nickel-based stainless steels, corrosion of welded joints can be severe, despite that they rarely undergo galvanic corrosion when mechanically joined.

Safety issues

Welding can be a dangerous and unhealthy practice without the proper precautions; however, with the use of new technology and proper protection the risks of injury or death associated with welding can be greatly reduced.

Heat and sparks

Because many common welding procedures involve an open electric arc or flame, the risk of burns from heat and sparks is significant. To prevent them, welders wear protective clothing in the form of heavy leather gloves and protective long sleeve jackets to avoid exposure to extreme heat, flames, and sparks.

Eye damage

The brightness of the weld area leads to a condition called arc eye in which ultraviolet light causes inflammation of the cornea and can burn the retinas of the eyes. Welding goggles and helmets with dark face plates are worn to prevent this exposure and, in recent years, new helmet models have been produced featuring a face plate that self-darkens upon exposure to high amounts of UV light.

To protect bystanders, transparent welding curtains often surround the welding area. These curtains, made of a polyvinyl chloride plastic film, shield nearby workers from exposure to the UV light from the electric arc, but should not be used to replace the filter glass used in helmets.

In 1970, a Swedish doctor, Åke Sandén, developed a new type of welding goggles that used a multilayer interference filter to block most of the light from the arc. He had observed that most welders could not see well enough, with the mask on, to strike the arc, so they would flip the mask up, then flip it down again once the arc was going: this exposed their naked eyes to the intense light for a while.

By coincidence, the spectrum of an electric arc has a notch in it, which coincides with the yellow sodium line. Thus, a welding shop could be lit by sodium vapor lamps or daylight, and the welder could see well to strike the arc. The Swedish government required these masks to be used for arc welding, but they were not used in the United States. They may have disappeared.

Inhaled matter

Welders are also often exposed to dangerous gases and particulate matter. Processes like flux-cored arc welding and shielded metal arc welding produce smoke containing particles of various types of oxides. The size of the particles in question tends to influence the toxicity of the fumes, with smaller particles presenting a greater danger.

Additionally, many processes produce various gases (most commonly carbon dioxide and ozone, but others as well) that can prove dangerous if ventilation is inadequate. Furthermore, the use of compressed gases and flames in many welding processes pose an explosion and fire risk; some common precautions include limiting the amount of oxygen in the air and keeping combustible materials away from the workplace.

Gas tungsten arc welding

It is an arc welding process that uses a nonconsumable tungsten electrode to produce the weld. The weld area is protected from atmospheric contamination by a shielding gas (usually an inert gas such as argon), and a filler metal is normally used, though some welds, known as autogenous welds, do not require it. A constant-current welding power supply produces energy which is conducted across the arc through a column of highly ionized gas and metal vapors known as a plasma.

GTAW is most commonly used to weld thin sections of stainless steel and non-ferrous metals such as aluminum, magnesium, and copper alloys. The process grants the operator greater control over the weld than competing processes such as shielded metal arc welding and gas metal arc welding, allowing for stronger, higher quality welds. However, GTAW is comparatively more complex and difficult to master, and furthermore, it is significantly slower than most other welding techniques. A related process, plasma arc welding, uses a slightly different welding torch to create a more focused welding arc and as a result is often automated.

Development

After the discovery of the electric arc in 1800 by Humphry Davy, arc welding developed slowly. C. L. Coffin had the idea of welding in an inert gas atmosphere in 1890, but even in the early 20th century, welding non-ferrous materials like aluminum and magnesium remained difficult, because these metals reacted rapidly with the air, resulting in porous and dross-filled welds. Processes using flux-covered electrodes did not satisfactorily protect the weld area from contamination. To solve the problem, bottled inert gases were used in the beginning of the 1930s. A few years later, a direct current, gas-shielded welding process emerged in the aircraft industry for welding magnesium.

This process was perfected in 1941, and became known as heliarc or tungsten inert gas welding, because it utilized a tungsten electrode and helium as a shielding gas. Initially, the electrode overheated quickly, and in spite of tungsten's high melting temperature, particles of tungsten were transferred to the weld. To address this problem, the polarity of the electrode was changed from positive to negative, but this made it unsuitable for welding many non-ferrous materials. Finally, the development of alternating current units made it possible to stabilize the arc and produce high quality aluminum and magnesium welds.

Developments continued during the following decades. Linde Air Products developed water-cooled torches that helped to prevent overheating when welding with high currents. Additionally, during the 1950s, as the process continued to gain popularity, some users turned to carbon dioxide as an alternative to the more expensive welding atmospheres consisting of argon and helium. However, this proved unacceptable for welding aluminum and magnesium because it reduced weld quality, and as a result, it is rarely used with GTAW today.

In 1953, a new process based on GTAW was developed, called plasma arc welding. It affords greater control and improves weld quality by using a nozzle to focus the electric arc, but is largely limited to automated systems, whereas GTAW remains primarily a manual, hand-held method. Development within the GTAW process has continued as well, and today a number of variations exist. Among the most popular are the pulsed-current, manual programmed, hot-wire, dabber, and increased penetration GTAW methods.

OPERATION

Manual gas tungsten arc welding is often considered the most difficult of all the welding processes commonly used in industry. Because the welder must maintain a short arc length, great care and skill are required to prevent contact between the electrode and the workpiece. Similar to torch welding, GTAW normally requires two hands, since most applications require that the welder manually feed a filler metal into the weld area with one hand while manipulating the welding torch in the other. However, some welds combining thin materials (known as autogenous or fusion welds) can be accomplished without filler metal; most notably edge, corner, and butt joints.

To strike the welding arc, a high frequency generator (similar to a Tesla coil) provides an electric spark; this spark is a conductive path for the welding current through the shielding gas and allows the arc to be initiated while the electrode and the workpiece are separated, typically about 1.5-3 mm (0.06-0.12 in) apart. This high voltage, high frequency burst can be damaging to some vehicle electrical systems and electronics, because induced voltages on vehicle wiring can also cause small conductive sparks in the vehicle wiring or within semiconductor packaging. Vehicle 12V power may conduct across these ionized paths, driven by the high-current 12V vehicle battery. These currents can be sufficiently destructive as to disable the vehicle; thus the warning to disconnect the vehicle battery power from both +12 and ground before using welding equipment on vehicles.

An alternate way to initiate the arc is the "scratch start". Scratching the electrode against the work with the power on also serve to strike an arc, in the same way as SMAW ("stick") arc welding. However, scratch starting can cause contamination of the weld and electrode. Some GTAW equipment is capable of a mode called "touch start" or "lift arc"; here the equipment reduces the voltage on the electrode to only a few volts, with a current limit of one or two amps (well below the limit that causes metal to transfer and contamination of the weld or electrode). When the GTAW equipment detects that the electrode has left the surface and a spark is present, it immediately (within microseconds) increases power, converting the spark to a full arc.

Once the arc is struck, the welder moves the torch in a small circle to create a welding pool, the size of which depends on the size of the electrode and the amount of current. While maintaining a constant separation between the electrode and the workpiece, the operator then moves the torch back slightly and tilts it backward about 10-15 degrees from vertical. Filler metal is added manually to the front end of the weld pool as it is needed.

Welders often develop a technique of rapidly alternating between moving the torch forward (to advance the weld pool) and adding filler metal. The filler rod is withdrawn from the weld pool each time the electrode advances, but it is never removed from the gas shield to prevent oxidation of its surface and contamination of the weld. Filler rods composed of metals with low melting temperature, such as aluminum, require that the operator maintain some distance from the arc while staying inside the gas shield.

If held too close to the arc, the filler rod can melt before it makes contact with the weld puddle. As the weld nears completion, the arc current is often gradually reduced to allow the weld crater to solidify and prevent the formation of crater cracks at the end of the weld.

Operation modes

GTAW can use a positive direct current, negative direct current or an alternating current, depending on the power supply set up. A negative direct current from the electrode causes a stream of electrons to collide with the surface, generating large amounts of heat at the weld region. This creates a deep, narrow weld. In the opposite process where the electrode is connected to the positive power supply terminal, electrons flow from the part being welded to the tip of the electrode instead, so the heating action of the electrons is mostly on the electrode. This mode also helps to remove oxide layers from the surface of the region to be welded, which is good for metals such as aluminum or magnesium. A shallow, wide weld is produced from this mode, with minimum heat input. Alternating current gives a combination of negative and positive modes, giving a cleaning effect and imparts a lot of heat as well.

Safety

Like other arc welding processes, GTAW can be dangerous if proper precautions are not taken. The process produces intense ultraviolet radiation, which can cause a form of sunburn and, in a few cases, trigger the development of skin cancer. Flying sparks and droplets of molten metal can cause severe burns and start a fire if flammable material is nearby, though GTAW generally produces very few sparks or metal droplets when performed properly.

It is essential that the welder wear suitable protective clothing, including leather gloves, a closed shirt collar to protect the neck (especially the throat), a protective long sleeve jacket and a suitable welding helmet to prevent retinal damage or ultraviolet burns to the cornea, often called arc eye. The shade of welding lens will depend upon the amperage of the welding current.

Due to the absence of smoke in GTAW, the arc appears brighter than shielded metal arc welding and more ultraviolet radiation is produced. Exposure of bare skin near a GTAW arc for even a few seconds may cause a painful sunburn. Additionally, the tungsten electrode is heated to a white hot state like the filament of a light bulb, adding greatly to the total radiated light and heat energy. Transparent welding curtains, made of a polyvinyl chloride plastic film, dyed in order to block UV radiation, are often used to shield nearby personnel from exposure.

Welders are also often exposed to dangerous gases and particulate matter. Shielding gases can displace oxygen and lead to asphyxiation, and while smoke is not produced, the arc in GTAW produces very short wavelength ultraviolet light, which causes surrounding air to break down and form ozone.

Metals will volatilize and heavy metals can be taken into the lungs. Similarly, the heat can cause poisonous fumes to form from cleaning and degreasing materials. For example, chlorinated products such as brake cleaner on a weld surface generate the toxic gas phosgene when heated. Cleaning operations using these agents should not be performed near the site of welding, and proper ventilation is necessary to protect the welder.

Applications

While the aerospace industry is one of the primary users of gas tungsten arc welding, the process is used in a number of other areas. Many industries use GTAW for welding thin workpieces, especially nonferrous metals. It is used extensively in the manufacture of space vehicles, and is also frequently employed to weld small-diameter, thin-wall tubing such as those used in the bicycle industry. In addition, GTAW is often used to make root or first pass welds for piping of various sizes.

In maintenance and repair work, the process is commonly used to repair tools and dies, especially components made of aluminum and magnesium. Because the weld metal is not transferred directly across the electric arc like most open arc welding processes, a vast assortment of welding filler metal is available to the welding engineer. In fact, no other welding process permits the welding of so many alloys in so many product configurations. Filler metal alloys, such as elemental aluminum and chromium, can be lost through the electric arc from volatilization. This loss does not occur with the GTAW process. Because the resulting welds have the same chemical integrity as the original base metal or match the base metals more closely, GTAW welds are highly resistant to corrosion and cracking over long time periods, GTAW is the welding procedure of choice for critical welding operations like sealing spent nuclear fuel canisters before burial.

Quality

Engineers prefer GTAW welds because of its low-hydrogen properties and the match of mechanical and chemical properties with the base material. Maximum weld quality is assured by maintaining the cleanliness of the operation-all equipment and materials used must be free from oil, moisture, dirt and other impurities, as these cause weld porosity and consequently a decrease in weld strength and quality.

To remove oil and grease, alcohol or similar commercial solvents may be used, while a stainless steel wire brush or chemical process can remove oxides from the surfaces of metals like aluminum. Rust on steels can be removed by first grit blasting the surface and then using a wire brush to remove any embedded grit. These steps are especially important when negative polarity direct current is used, because such a power supply provides no cleaning during the welding process, unlike positive polarity direct current or alternating current.

To maintain a clean weld pool during welding, the shielding gas flow should be sufficient and consistent so that the gas covers the weld and blocks impurities in the atmosphere. GTA welding in windy or drafty environments increases the amount of shielding gas necessary to protect the weld, increasing the cost and making the process unpopular outdoors.

Because of GTAW's relative difficulty and the importance of proper technique, skilled operators are employed for important applications. Welders in the U.S. should be qualified following the requirements of the American Welding Society or American Society of Mechanical Engineers. Low heat input, caused by low welding current or high welding speed, can limit penetration and cause the weld bead to lift away from the surface being welded. If there is too much heat input, however, the weld bead grows in width while the likelihood of excessive penetration and spatter increase. Additionally, if the welding torch is too far from the workpiece the shielding gas becomes ineffective causing porosity within the weld. This results in a weld with pinholes, which is weaker than a typical weld.

If the amount of current used exceeds the capability of the electrode, tungsten inclusions in the weld may result. Known as tungsten spitting, it can be identified with radiography and prevented by changing the type of electrode or increasing the electrode diameter. In addition, if the electrode is not well protected by the gas shield or the operator accidentally allows it to contact the molten metal, it can become dirty or contaminated. This often causes the welding arc to become unstable, requiring that electrode be ground with a diamond abrasive to remove the impurity.

The equipment required for the gas tungsten arc welding operation includes a welding torch utilizing a nonconsumable tungsten electrode, a constant-current welding power supply, and a shielding gas source.

Welding torch

GTAW welding torches are designed for either automatic or manual operation and are equipped with cooling systems using air or water. The automatic and manual torches are similar in construction, but the manual torch has a handle while the automatic torch normally comes with a mounting rack.

The angle between the centerline of the handle and the centerline of the tungsten electrode, known as the head angle, can be varied on some manual torches according to the preference of the operator. Air cooling systems are most often used for low-current operations (up to about 200 A), while water cooling is required for high-current welding (up to about 600 A). The torches are connected with cables to the power supply and with hoses to the shielding gas source and where used, the water supply.

The internal metal parts of a torch are made of hard alloys of copper or brass in order to transmit current and heat effectively. The tungsten electrode must be held firmly in the center of the torch with an appropriately sized collet, and ports around the electrode provide a constant flow of shielding gas. Collets are sized according to the diameter of the tungsten electrode they hold. The body of the torch is made of heat-resistant, insulating plastics covering the metal components, providing insulation from heat and electricity to protect the welder.

The size of the welding torch nozzle depends on the amount of shielded area desired. The size of the gas nozzle will depend upon the diameter of the electrode, the joint configuration, and the availability of access to the joint by the welder. The inside diameter of the nozzle is preferably at least three times the diameter of the electrode, but there are no hard rules. The welder will judge the effectiveness of the shielding and increase the nozzle size to increase the area protected by the external gas shield as needed.

The nozzle must be heat resistant and thus is normally made of alumina or a ceramic material, but fused quartz, a glass-like substance, offers greater visibility. Devices can be inserted into the nozzle for special applications, such as gas lenses or valves to improve the control shielding gas flow to reduce turbulence and introduction of contaminated atmosphere into the shielded area. Hand switches to control welding current can be added to the manual GTAW torches.

Power supply

Gas tungsten arc welding uses a constant current power source, meaning that the current (and thus the heat) remains relatively constant, even if the arc distance and voltage change. This is important because most applications of GTAW are manual or semiautomatic, requiring that an operator hold the torch. Maintaining a suitably steady arc distance is difficult if a constant voltage power source is used instead, since it can cause dramatic heat variations and make welding more difficult.

The preferred polarity of the GTAW system depends largely on the type of metal being welded. Direct current with a negatively charged electrode (DCEN) is often employed when welding steels, nickel, titanium, and other metals. It can also be used in automatic GTA welding of aluminum or magnesium when helium is used as a shielding gas.

The negatively charged electrode generates heat by emitting electrons which travel across the arc, causing thermal ionization of the shielding gas and increasing the temperature of the base material. The ionized shielding gas flows toward the electrode, not the base material. Direct current with a positively charged electrode (DCEP) is less common, and is used primarily for shallow welds since less heat is generated in the base material.

Instead of flowing from the electrode to the base material, as in DCEN, electrons go the other direction, causing the electrode to reach very high temperatures. To help it maintain its shape and prevent softening, a larger electrode is often used. As the electrons flow toward the electrode, ionized shielding gas flows back toward the base material, cleaning the weld by removing oxides and other impurities and thereby improving its quality and appearance.

Alternating current, commonly used when welding aluminum and magnesium manually or semi-automatically, combines the two direct currents by making the electrode and base material alternate between positive and negative charge. This causes the electron flow to switch directions constantly, preventing the tungsten electrode from overheating while maintaining the heat in the base material. Surface oxides are still removed during the electrode-positive portion of the cycle and the base metal is heated more deeply during the electrode-negative portion of the cycle.

Some power supplies enable operators to use an unbalanced alternating current wave by modifying the exact percentage of time that the current spends in each state of polarity, giving them more control over the amount of heat and cleaning action supplied by the power source. In addition, operators must be wary of rectification, in which the arc fails to reignite as it passes from straight polarity (negative electrode) to reverse polarity (positive electrode). To remedy the problem, a square wave power supply can be used, as can high-frequency voltage to encourage ignition.

The electrode used in GTAW is made of tungsten or a tungsten alloy, because tungsten has the highest melting temperature among pure metals, at 3,422 °C (6,192 °F). As a result, the electrode is not consumed during welding, though some erosion (called burn-off) can occur. Electrodes can have either a clean finish or a ground finish-clean finish electrodes have been chemically cleaned, while ground finish electrodes have been ground to a uniform size and have a polished surface, making them optimal for heat conduction. The diameter of the electrode can vary between 0.5 and 6.4 millimetres (0.02 and 0.25 in), and their length can range from 75 to 610 millimetres (3.0 to 24 in).

A number of tungsten alloys have been standardized by the International Organization for Standardization and the American Welding Society in ISO 6848 and AWS A5.12, respectively, for use in GTAW electrodes, and are summarized in the adjacent table. Pure tungsten electrodes (classified as WP or EWP) are general purpose and low cost electrodes. Cerium oxide (or ceria) as an alloying element improves arc stability and ease of starting while decreasing burn-off. Using an alloy of lanthanum oxide (or lanthana) has a similar effect. Thorium oxide (or thoria) alloy electrodes were designed for DC applications and can withstand somewhat higher temperatures while providing many of the benefits of other alloys. However, it is somewhat radioactive.

Inhalation of the thorium grinding dust during preparation of the electrode is hazardous to one's health. As a replacement to thoriated electrodes, electrodes with larger concentrations of lanthanum oxide can be used. Electrodes containing zirconium oxide (or zirconia) increase the current capacity while improving arc stability and starting and increasing electrode life. In addition, electrode manufacturers may create alternative tungsten alloys with specified metal additions, and these are designated with the classification EWG under the AWS system.

Filler metals are also used in nearly all applications of GTAW, the major exception being the welding of thin materials. Filler metals are available with different diameters and are made of a variety of materials. In most cases, the filler metal in the form of a rod is added to the weld pool manually, but some applications call for an automatically fed filler metal, which often is stored on spools or coils.

Shielding gas

As with other welding processes such as gas metal arc welding, shielding gases are necessary in GTAW to protect the welding area from atmospheric gases such as nitrogen and oxygen, which can cause fusion defects, porosity, and weld metal embrittlement if they come in contact with the electrode, the arc, or the welding metal. The gas also transfers heat from the tungsten electrode to the metal, and it helps start and maintain a stable arc.

The selection of a shielding gas depends on several factors, including the type of material being welded, joint design, and desired final weld appearance. Argon is the most commonly used shielding gas for GTAW, since it helps prevent defects due to a varying arc length. When used with alternating current, the use of argon results in high weld quality and good appearance.

Another common shielding gas, helium, is most often used to increase the weld penetration in a joint, to increase the welding speed, and to weld metals with high heat conductivity, such as copper and aluminum. A significant disadvantage is the difficulty of striking an arc with helium gas, and the decreased weld quality associated with a varying arc length.

Argon-helium mixtures are also frequently utilized in GTAW, since they can increase control of the heat input while maintaining the benefits of using argon. Normally, the mixtures are made with primarily helium (often about 75% or higher) and a balance of argon. These mixtures increase the speed and quality of the AC welding of aluminum, and also make it easier to strike an arc.

Another shielding gas mixture, argon-hydrogen, is used in the mechanized welding of light gauge stainless steel, but because hydrogen can cause porosity, its uses are limited. Similarly, nitrogen can sometimes be added to argon to help stabilize the austenite in austentitic stainless steels and increase penetration when welding copper. Due to porosity problems in ferritic steels and limited benefits, however, it is not a popular shielding gas additive.

Materials

Gas tungsten arc welding is most commonly used to weld stainless steel and nonferrous materials, such as aluminum and magnesium, but it can be applied to nearly all metals, with notable exceptions being lead and zinc. Its applications involving carbon steels are limited not because of process restrictions, but because of the existence of more economical steel welding techniques, such as gas metal arc welding and shielded metal arc welding. Furthermore, GTAW can be performed in a variety of other-than-flat positions, depending on the skill of the welder and the materials being welded.

Aluminum and magnesium

Aluminum and magnesium are most often welded using alternating current, but the use of direct current is also possible, depending on the properties desired. Before welding, the work area should be cleaned and may be preheated to 175 to 200 °C (347 to 392 °F) for aluminum or to a maximum of 150 °C (302 °F) for thick magnesium workpieces to improve penetration and increase travel speed. AC current can provide a self-cleaning effect, removing the thin, refractory aluminium oxide (sapphire) layer that forms on aluminium metal within minutes of exposure to air.

This oxide layer must be removed for welding to occur. When alternating current is used, pure tungsten electrodes or zirconiated tungsten electrodes are preferred over thoriated electrodes, as the latter are more likely to "spit" electrode particles across the welding arc into the weld. Blunt electrode tips are preferred, and pure argon shielding gas should be employed for thin workpieces. Introducing helium allows for greater penetration in thicker workpieces, but can make arc starting difficult.

Direct current of either polarity, positive or negative, can be used to weld aluminum and magnesium as well. Direct current with a negatively charged electrode (DCEN) allows for high penetration. Argon is commonly used as a shielding gas for DCEN welding of aluminum.

Shielding gases with high helium contents are often used for higher penetration in thicker materials. Thoriated electrodes are suitable for use in DCEN welding of aluminum. Direct current with a positively charged electrode (DCEP) is used primarily for shallow welds, especially those with a joint thickness of less than 1.6 mm (0.063 in). A thoriated tungsten electrode is commonly used, along with a pure argon shielding gas.

STEELS

For GTA welding of carbon and stainless steels, the selection of a filler material is important to prevent excessive porosity. Oxides on the filler material and workpieces must be removed before welding to prevent contamination, and immediately prior to welding, alcohol or acetone should be used to clean the surface.

Preheating is generally not necessary for mild steels less than one inch thick, but low alloy steels may require preheating to slow the cooling process and prevent the formation of martensite in the heat-affected zone. Tool steels should also be preheated to prevent cracking in the heat-affected zone. Austenitic stainless steels do not require preheating, but martensitic and ferritic chromium stainless steels do. A DCEN power source is normally used, and thoriated electrodes, tapered to a sharp point, are recommended. Pure argon is used for thin workpieces, but helium can be introduced as thickness increases.

Copper alloys

TIG welding of copper and some of its alloys is possible, but in order to get a seam free of oxidation and porosities, shielding gas needs to be provided on the root side of the weld. Alternatively, a special "backing tape", consisting of a fiberglass weave on heat-resistant aluminum tape can be used, to prevent air reaching the molten metal.

Dissimilar metals

Welding dissimilar metals often introduces new difficulties to GTAW welding, because most materials do not easily fuse to form a strong bond. However, welds of dissimilar materials have numerous applications in manufacturing, repair work, and the prevention of corrosion and oxidation. In some joints, a compatible filler metal is chosen to help form the bond, and this filler metal can be the same as one of the base materials (for example, using a stainless steel filler metal with stainless steel and carbon steel as base materials), or a different metal (such as the use of a nickel filler metal for joining steel and cast iron). Very different materials may be coated or "buttered" with a material compatible with a particular filler metal, and then welded. In addition, GTAW can be used in cladding or overlaying dissimilar materials.

When welding dissimilar metals, the joint must have an accurate fit, with proper gap dimensions and bevel angles. Care should be taken to avoid melting excessive base material. Pulsed current is particularly useful for these applications, as it helps limit the heat input. The filler metal should be added quickly, and a large weld pool should be avoided to prevent dilution of the base materials.

Process variations

Pulsed-current

In the pulsed-current mode, the welding current rapidly alternates between two levels. The higher current state is known as the pulse current, while the lower current level is called the background current. During the period of pulse current, the weld area is heated and fusion occurs. Upon dropping to the background current, the weld area is allowed to cool and solidify. Pulsed-current GTAW has a number of advantages, including lower heat input and consequently a reduction in distortion and warpage in thin workpieces. In addition, it allows for greater control of the weld pool, and can increase weld penetration, welding speed, and quality. A similar method, manual programmed GTAW, allows the operator to program a specific rate and magnitude of current variations, making it useful for specialized applications.

Dabber

The dabber variation is used to precisely place weld metal on thin edges. The automatic process replicates the motions of manual welding by feeding a cold filler wire into the weld area and dabbing (or oscillating) it into the welding arc. It can be used in conjunction with pulsed current, and is used to weld a variety of alloys, including titanium, nickel, and tool steels. Common applications include rebuilding seals in jet engines and building up saw blades, milling cutters, drill bits, and mower blades.

Hot wire

Welding filler metal can be resistance heated to a temperature near its melting point before being introduced into the weld pool. This increases the deposition rate of machine and automatic GTAW welding processes. More pounds per hour of filler metal is introduced into the weld joint than when filler metal is added cold and the heat of the electric arc introduces all of the heat. This process is used extensively in base material build up before machining, clad metal overlays, and hardfacing operations.

6

BOILER AND ACCESSORIES

A boiler is a closed vessel in which water or other fluid is heated. The heated or vaporized fluid exits the boiler for use in various processes or heating applications.The pressure vessel in a boiler is usually made of steel (or alloy steel), or historically of wrought iron. Stainless steel is virtually prohibited (by the ASME Boiler Code) for use in wetted parts of modern boilers, but is used often in superheater sections that will not be exposed to liquid boiler water. In live steam models, copper or brass is often used because it is more easily fabricated in smaller size boilers.

Historically, copper was often used for fireboxes (particularly for steam locomotives), because of its better formability and higher thermal conductivity; however, in more recent times, the high price of copper often makes this an uneconomic choice and cheaper substitutes (such as steel) are used instead.

For much of the Victorian "age of steam", the only material used for boilermaking was the highest grade of wrought iron, with assembly by rivetting. This iron was often obtained from specialist ironworks, such as at Cleator Moor (UK), noted for the high quality of their rolled plate and its suitability for high-reliability use in critical applications, such as high-pressure boilers. In the 20th century, design practice instead moved towards the use of steel, which is stronger and cheaper, with welded construction, which is quicker and requires less labour.

Cast iron may be used for the heating vessel of domestic water heaters. Although such heaters are usually termed "boilers" in some countries, their purpose is usually to produce hot water, not steam, and so they run at low pressure and try to avoid actual boiling. The brittleness of cast iron makes it impractical for high pressure steam boilers.

FUEL

The source of heat for a boiler is combustion of any of several fuels, such as wood, coal, oil, or natural gas. Electric steam boilers use resistance- or immersion-type heating elements. Nuclear fission is also used as a heat source for generating steam. Heat recovery steam generators (HRSGs) use the heat rejected from other processes such as gas turbines.

Configurations

Boilers can be classified into the following configurations:

- "Pot boiler" or "Haycock boiler": a primitive "kettle" where a fire heats a partially-filled water container from below. 18th century Haycock boilers generally produced and stored large volumes of very low-pressure steam, often hardly above that of the atmosphere. These could burn wood or most often, coal. Efficiency was very low.
- Fire-tube boiler. Here, water partially fills a boiler barrel with a small volume left above to accommodate the steam (steam space). This is the type of boiler used in nearly all steam locomotives. The heat source is inside a furnace or firebox that has to be kept permanently surrounded by the water in order to maintain the temperature of the heating surface just below boiling point. The furnace can be situated at one end of a fire-tube which lengthens the path of the hot gases, thus augmenting the heating surface which can be further increased by making the gases reverse direction through a second parallel tube or a bundle of multiple tubes (two-pass or return flue boiler); alternatively the gases may be taken along the sides and then beneath the boiler through flues (3-pass boiler). In the case of a locomotive-type boiler, a boiler barrel extends from the firebox and the hot gases pass through a bundle of fire tubes inside the barrel which greatly increase the heating surface compared to a single tube and further improve heat transfer. Fire-tube boilers usually have a comparatively low rate of steam production, but high steam storage capacity. Fire-tube boilers mostly burn solid fuels, but are readily adaptable to those of the liquid or gas variety.
- Water-tube boiler. In this type,the water tubes are arranged inside a furnace in a number of possible configurations: often the water tubes connect large drums, the lower ones containing water and the upper ones, steam and water; in other cases, such as a monotube boiler, water is circulated by a pump through a succession of coils. This type generally gives high steam production rates, but less storage capacity than the above. Water

tube boilers can be designed to exploit any heat source and are generally preferred in high pressure applications since the high pressure water/steam is contained within small diameter pipes which can withstand the pressure with a thinner wall.

- Flash boiler. A specialized type of water-tube boiler.
- Fire-tube boiler with Water-tube firebox. Sometimes the two above types have been combined in the following manner: the firebox contains an assembly of water tubes, called thermic syphons. The gases then pass through a conventional firetube boiler. Water-tube fireboxes were installed in many Hungarian locomotives, but have met with little success in other countries.
- Sectional boiler. In a cast iron sectional boiler, sometimes called a "pork chop boiler" the water is contained inside cast iron sections. These sections are assembled on site to create the finished boiler.

Safety

Historically, boilers were a source of many serious injuries and property destruction due to poorly understood engineering principles. Thin and brittle metal shells can rupture, while poorly welded or riveted seams could open up, leading to a violent eruption of the pressurized steam. Collapsed or dislodged boiler tubes could also spray scalding-hot steam and smoke out of the air intake and firing chute, injuring the firemen who loaded coal into the fire chamber. Extremely large boilers providing hundreds of horsepower to operate factories could demolish entire buildings.

A boiler that has a loss of feed water and is permitted to boil dry can be extremely dangerous. If feed water is then sent into the empty boiler, the small cascade of incoming water instantly boils on contact with the superheated metal shell and leads to a violent explosion that cannot be controlled even by safety steam valves. Draining of the boiler could also occur if a leak occurred in the steam supply lines that was larger than the make-up water supply could replace. The Hartford Loop was invented in 1919 by the Hartford Steam Boiler and Insurance Company as a method to help prevent this condition from occurring, and thereby reduce their insurance claims.

Super-heater

A superheater is a device used to convert saturated steam or wet steam into dry steam used for power generation or processes. There are three types of superheaters namely: radiant, convection, and separately fired. A superheater can vary in size from a few tens of feet to several hundred feet (a few meters or some hundred meters).

(a) A radiant superheater is placed directly in the combustion chamber.

(b) A convection superheater is located in the path of the hot gases.

(c) A separately fired superheater, as its name implies, is totally separated from the boiler.

A superheater is a device in a steam engine, when considering locomotives, that heats the steam generated by the boiler again, increasing its thermal energy and decreasing the likelihood that it will condense inside the engine. Superheaters increase the efficiency of the steam engine, and were widely adopted. Steam which has been superheated is logically known as superheated steam; non-superheated steam is called saturated steam or wet steam. Superheaters were applied to steam locomotives in quantity from the early 20th century, to most steam vehicles, and to stationary steam engines. This equipment is still an integral part of power generating stations throughout the world.

LOCOMOTIVE USE

In steam locomotive use, by far the most common form of superheater is the fire-tube type. This takes the saturated steam supplied in the dry pipe into a superheater header mounted against the tube sheet in the smokebox. The steam is then passed through a number of superheater elements-long pipes which are placed inside special, widened fire tubes, called flues.

Hot combustion gases from the locomotive's fire pass through these flues just like they do the firetubes, and as well as heating the water they also heat the steam inside the superheater elements they flow over. The super-heater element doubles back on itself so that the heated steam can return; most do this twice at the fire end and once at the smokebox end, so that the steam travels a distance of four times the header's length while being heated. The superheated steam, at the end of its journey through the elements, passes into a separate compartment of the superheater header and then to the cylinders as normal.

Damper and snifting valve

The steam passing through the superheater elements cools their metal and prevents them from melting, but when the throttle closes this cooling effect is absent, and thus a damper closes in the smokebox to cut off the flow through the flues and prevent them being damaged. Some locomotives (particularly on the London and

North Eastern Railway) were fitted with snifting valves which admitted air to the superheater when the locomotive was coasting. This kept the superheater elements cool and the cylinders warm. The snifting valve can be seen behind the chimney on many LNER locomotives.

Front-end throttle

A superheater increases the distance between the throttle and the cylinders in the steam circuit and thus reduces the immediacy of throttle action. To counteract this, some later steam locomotives were fitted with a front-end throttle-placed in the smokebox after the superheater. Such locomotives can sometimes be identified by an external throttle rod that stretches the whole length of the boiler, with a crank on the outside of the smokebox. This arrangement also allows superheated steam to be used for auxiliary appliances, such as the dynamo and air pumps.

Another benefit of the front end throttle is that superheated steam is immediately available. With the dome throttle it took quite some time before the super heater actually provided benefits in efficiency. One can think of it in this way: if one opens saturated steam from the boiler to the super-heater it goes straight through the super-heater units and to the cylinders which doesn't leave much time for the steam to be superheated. With the front-end throttle, steam is in the super-heater units while the engine is sitting at the station and that steam is being superheated. Then when the throttle is opened, superheated steam goes to the cylinders immediately.

Cylinder valves

Locomotives with super-heaters are usually fitted with piston valves or poppet valves. This is because it is difficult to keep a slide valve properly lubricated at high temperature.

Applications

The first practical super-heater was developed in Germany by Wilhelm Schmidt during the 1880s and 1890s, and the benefits of the invention were demonstrated in the U.K. by the Great Western Railway in 1906. The G.W.R. Chief Mechanical Engineer, G. J. Churchward believed, however, that the Schmidt type could be bettered, and design and testing of an indigenous Swindon type was undertaken, culminating in the Standard Type 3 in 1909. Douglas Earle Marsh carried out a series of comparative tests between members of his I3 class using saturated steam and those fitted with the Schmidt super-heater between October 1907 and March 1910, proving the advantages of the latter in terms of performance and efficiency.

Other improved super-heaters were introduced by John G. Robinson of the Great Central Railway at Gorton locomotive works, by Robert Urie of the London and South Western Railway (LSWR) at Eastleigh railway works, and Richard Maunsell of the Southern Railway (Great Britain), also at Eastleigh.

URIE'S "EASTLEIGH" SUPERHEATER

Robert Urie's design of superheater for the LSWR was the product of experience with his H15 class 4-6-0 locomotives. In anticipation of performance trials, eight examples were fitted with Schmidt and Robinson superheaters, and two others remained saturated. However, the First World War intervened before the trials could take place, although an LSWR Locomotive Committee report from late 1915 noted that the Robinson version gave the best fuel efficiency. It gave an average of 48.35 lb (21.9 kg) coal consumed per mile over an average distance of 39,824 mi (64,090.5 km), compared to 48.42 lb (22.0 kg) and 59.05 lb (26.8 kg) coal for the Schmidt and saturated examples respectively.

However, the report stated that both superheater types had serious drawbacks, with the Schmidt system featuring a damper control on the superheater header that caused hot gases to condense into sulphuric acid, which caused pitting and subsequent weakening of the superheater elements.

Leakage of gases was also commonplace between the elements and the header, and maintenance was difficult without removal of the horizontally-arranged assembly. The Robinson version suffered from temperature variations caused by saturated and superheated steam chambers being adjacent, causing material stress, and had similar access problems as the Schmidt type.

The report's recommendations enabled Urie to design a new type of superheater with separate saturated steam headers above and below the superheater header. These were connected up by elements beginning at the saturated header, running through the flue tubes and back to the superheater header, and the whole assembly was vertically arranged for ease of maintenance. The device was highly successful in service, but was heavy and expensive to construct.

Advantages and disadvantages

The main advantages of using a superheater are reduced fuel and water consumption but there is a price to pay in increased maintenance costs. In most cases the benefits outweighed the costs and superheaters were widely used. An exception was shunting locomotives (switchers). British shunting locomotives were rarely fitted with superheaters. In locomotives used for mineral traffic the advantages seem to have been marginal. For example, the North Eastern Railway fitted superheaters to some of its NER Class P mineral locomotives but later began to remove them.

Without careful maintenance superheaters are prone to a particular type of hazardous failure in the tube bursting at the U-shaped turns in the superheater tube. This is difficult to both manufacture, and test when installed, and a rupture will cause the superheated high-pressure steam to escape immediately into the large flues, and then back to the fire and into the cab, to the extreme danger of the locomotive crew.

Supercritical steam generators

Supercritical steam generators (also known as Benson boilers) are frequently used for the production of electric power. They operate at "supercritical pressure". In contrast to a "subcritical boiler", a supercritical steam generator operates at such a high pressure (over 3,200 psi/22.06 MPa or 220.6 bar) that actual boiling ceases to occur, and the boiler has no water - steam separation. There is no generation of steam bubbles within the water, because the pressure is above the "critical pressure" at which steam bubbles can form. It passes below the critical point as it does work in the high pressure turbine and enters the generator's condenser. This is more efficient, resulting in slightly less fuel use. The term "boiler" should not be used for a supercritical pressure steam generator, as no "boiling" actually occurs in this device.

History of supercritical steam generation

Contemporary supercritical steam generators are sometimes referred as Benson boilers. In 1922, Mark Benson was granted a patent for a boiler designed to convert water into steam at high pressure.

Safety was the main concern behind Benson's concept. Earlier steam generators were designed for relatively low pressures of up to about 100 bar (10,000 kPa; 1,450 psi), corresponding to the state of the art in steam turbine development at the time. One of their distinguishing technical characteristics was the riveted water/steam separator drum. These drums were where the water filled tubes were terminated after having passed through the boiler furnace.

These header drums were intended to be partially filled with water and above the water there was a baffle filled space where the boiler's steam and water vapour collected. The entrained water droplets were collected by the baffles and returned to the water pan. The mostly dry steam was piped out of the drum as the separated steam output of the boiler. These drums were often the source of boiler explosions, usually with catastrophic consequences.

However, this drum could be completely eliminated if the evaporation separation process was avoided altogether. This would happen if water entered the boiler at a pressure above the critical pressure (3,206 psi); was heated to a temperature above the critical temperature (706 degrees F) and then expanded (through a simple nozzle) to dry steam at some lower subcritical pressure. This could be obtained at a throttle valve located downstream of the evaporator section of the boiler.

As development of Benson technology continued, boiler design soon moved away from the original concept introduced by Mark Benson. In 1929, a test boiler that had been built in 1927 began operating in the thermal power plant at Gartenfeld in Berlin for the first time in subcritical mode with a fully open throttle valve. The second Benson boiler began operation in 1930 without a pressurizing valve at pressures between 40 and 180 bar (4,000 and 18,000 kPa; 580 and 2,611 psi) at the Berlin cable factory. This application represented the birth of the modern variable-pressure Benson boiler. After that development, the original patent was no longer used. The Benson boiler name, however, was retained.

Two current innovations have a good chance of winning acceptance in the competitive market for once-through steam generators:

- A new type of heat-recovery steam generator based on the Benson boiler, which has operated successfully at the Cottam combined-cycle power plant in the central part of England,
- The vertical tubing in the combustion chamber walls of coal-fired steam generators which combines the operating advantages of the Benson system with the design advantages of the drum-type boiler. Construction of a first reference plant, the Yaomeng power plant in China, commenced in 2001.

HYDRONIC BOILERS

Hydronic boilers are used in generating heat for residential and industrial purposes. They are the typical power plant for central heating systems fitted to houses in northern Europe (where they are commonly combined with domestic water heating), as opposed to the forced-air furnaces or wood burning stoves more common in North America. The hydronic boiler operates by way of heating water/fluid to a preset temperature (or sometimes in the case of single pipe systems, until it boils and turns to steam) and circulating that fluid throughout the home typically by way of radiators, baseboard heaters or through the floors. The fluid can be heated by any means...gas, wood, fuel oil, etc., but in built-up areas where piped gas is available, natural gas is currently the most economical and therefore the usual choice. The fluid is in an enclosed system and circulated throughout by means of a motorized pump. The name "boiler" can be a misnomer in that, except for systems using steam radiators, the water in a properly functioning hydronic boiler never actually boils. Most new systems are fitted with condensing boilers for greater efficiency. These boilers are referred to as condensing boilers because they condense the water vapor in the flue gases to capture the latent heat of vaporization of the water produced during combustion.

Hydronic systems are being used more and more in new construction in North America for several reasons. Among the reasons are:

- They are more efficient and more economical than forced-air systems (although initial installation can be more expensive, because of the cost of the copper and aluminum).
- The baseboard copper pipes and aluminum fins take up less room and use less metal than the bulky steel ductwork required for forced-air systems.
- They provide more even, less fluctuating temperatures than forced-air systems. The copper baseboard pipes hold and release heat over a longer period of time than air does, so the furnace does not have to switch off and on as much. (Copper heats mostly through conduction and radiation, whereas forced-air heats mostly through forced convection. Air has much lower thermal conductivity and volumetric heat capacity than copper, so the conditioned space warms up and cools down more quickly than with hydronic. See also thermal mass.)
- They tend to not dry out the interior air as much as forced air systems, but this is not always true. When forced air duct systems are air-sealed properly, and have return-air paths back to the furnace (thus reducing pressure differentials and therefore air movement between inside and outside the house), this is not an issue.
- They do not introduce any dust, allergens, mold, or (in the case of a faulty heat exchanger) combustion byproducts into the living space.

Forced-air heating does have some advantages, however. See forced-air heating.

Accessories

Boiler fittings and accessories

- Safety valve: It is used to relieve pressure and prevent possible explosion of a boiler.

- Water level indicators: They show the operator the level of fluid in the boiler, also known as a sight glass, water gauge or water column is provided.
- Bottom blowdown valves: They provide a means for removing solid particulates that condense and lie on the bottom of a boiler. As the name implies, this valve is usually located directly on the bottom of the boiler, and is occasionally opened to use the pressure in the boiler to push these particulates out.
- Continuous blowdown valve: This allows a small quantity of water to escape continuously. Its purpose is to prevent the water in the boiler becoming saturated with dissolved salts. Saturation would lead to foaming and cause water droplets to be carried over with the steam - a condition known as priming. Blowdown is also often used to monitor the chemistry of the boiler water.
- Flash Tank: High pressure blowdown enters this vessel where the steam can 'flash' safely and be used in a low-pressure system or be vented to atmosphere while the ambient pressure blowdown flows to drain.
- Automatic Blowdown/Continuous Heat Recovery System: This system allows the boiler to blowdown only when makeup water is flowing to the boiler, thereby transferring the maximum amount of heat possible from the blowdown to the makeup water. No flash tank is generally needed as the blowdown discharged is close to the temperature of the makeup water.
- Hand holes: They are steel plates installed in openings in "header" to allow for inspections & installation of tubes and inspection of internal surfaces.
- Steam drum internals, A series of screen, scrubber & cans (cyclone separators).
- Low- water cutoff: It is a mechanical means (usually a float switch) that is used to turn off the burner or shut off fuel to the boiler to prevent it from running once the water goes below a certain point. If a boiler is "dry-fired" (burned without water in it) it can cause rupture or catastrophic failure.
- Surface blowdown line: It provides a means for removing foam or other lightweight non-condensible substances that tend to float on top of the water inside the boiler.
- Circulating pump: It is designed to circulate water back to the boiler after it has expelled some of its heat.

- Feedwater check valve or clack valve: A non-return stop valve in the feedwater line. This may be fitted to the side of the boiler, just below the water level, or to the top of the boiler.
- Top feed: A check valve (clack valve) in the feedwater line, mounted on top of the boiler. It is intended to reduce the nuisance of limescale. It does not prevent limescale formation but causes the limescale to be precipitated in a powdery form which is easily washed out of the boiler.
- Desuperheater tubes or bundles: A series of tubes or bundles of tubes in the water drum or the steam drum designed to cool superheated steam. Thus is to supply auxiliary equipment that does not need, or may be damaged by, dry steam.
- Chemical injection line: A connection to add chemicals for controlling feedwater pH.

Steam accessories

- Main steam stop valve:
- Steam traps:
- Main steam stop/Check valve: It is used on multiple boiler installations.

Combustion accessories

- Fuel oil system:
- Gas system:
- Coal system:
- Soot blower

Other essential items

- Pressure gauges:
- Feed pumps:
- Fusible plug:
- Inspectors test pressure gauge attachment:
- Name plate:
- Registration plate:

CONTROLLING DRAUGHT

Most boilers now depend on mechanical draught equipment rather than natural draught. This is because natural draught is subject to outside air conditions and temperature of flue gases leaving the furnace, as well as the chimney height. All these factors make proper draught hard to attain and therefore make mechanical draught equipment much more economical.

There are three types of mechanical draught:

- Induced draught: This is obtained one of three ways, the first being the "stack effect" of a heated chimney, in which the flue gas is less dense than the ambient air surrounding the boiler. The denser column of ambient air forces combustion air into and through the boiler. The second method is through use of a steam jet. The steam jet oriented in the direction of flue gas flow induces flue gasses into the stack and allows for a greater flue gas velocity increasing the overall draught in the furnace. This method was common on steam driven locomotives which could not have tall chimneys. The third method is by simply using an induced draught fan (ID fan) which removes flue gases from the furnace and forces the exhaust gas up the stack. Almost all induced draught furnaces operate with a slightly negative pressure.
- Forced draught: Draught is obtained by forcing air into the furnace by means of a fan (FD fan) and ductwork. Air is often passed through an air heater; which, as the name suggests, heats the air going into the furnace in order to increase the overall efficiency of the boiler. Dampers are used to control the quantity of air admitted to the furnace. Forced draught furnaces usually have a positive pressure.
- Balanced draught: Balanced draught is obtained through use of both induced and forced draught. This is more common with larger boilers where the flue gases have to travel a long distance through many boiler passes. The induced draught fan works in conjunction with the forced draught fan allowing the furnace pressure to be maintained slightly below atmospheric.

Dealkalization of water

The dealkalization of water refers to the removal of alkalinity ions from water. Chloride cycle anion ion exchange dealkalizers remove alkalinity from water. Chloride cycle dealkalizers operate similar to sodium cycle cation water softeners. Like water softeners, dealkalizers contain ion exchange resins that are regenerated with a concentrated salt (brine) solution - NaCl. In the case of a water softener, the cation exchange resin is exchanging sodium (the Na ion of NaCl) for hardness minerals such as calcium and magnesium.

A dealkalizer contains strong base anion exchange resin that exchanges chloride (the Cl ion of the NaCl) for carbonate, bicarbonate and sulfate. As water passes through the anion resin the carbonate, bicarbonate and sulfate ions are exchanged for chloride ions.

"Higher capacities can be realized by use of type II rather than type I strong base anion resins. Although bicarbonates are not held as tightly as chlorides on the SBA (strong base anion) resins in the hydroxide form, when the resin is predominantly in the chloride form the pH has been raised by a small addition of caustic to the brine regenerant, there will be a favorable exchange of bicarbonate for the chloride.

This exchange works well only with high alkalinity waters (40% to 80%), with capacities of 4 to 10 Kg/CF being obtained. The advantages of SBA resin dealkalization is that low-cost salt is used in place of the acid necessary for the SAC (strong acid cation) and un-lined steel tanks can be used."

Why dealkalize?

Dealkalizers are most often used as pre-treatment to a boiler and are usually preceded by a water softener. Alkalinity is a factor that most often dictates the amount of boiler blowdown. High alkalinity promotes boiler foaming and carryover and causes high amounts of boiler blowoff. When alkalinity is the limiting factor affecting the amount of blowdown, a dealkalizer will increase the cycles of concentrations and reduce blowdown and operating costs.

The reduction of blowdown by dealkalization keeps the water treatment chemicals in the boiler longer, thus minimizing the amount of chemicals required for efficient, noncorrosive operation.

Carbonate and bicarbonate alkalinities are decomposed by heat in boiler water releasing carbon dioxide into the steam. This gas combines with the condensed steam in process equipment and return lines to form carbonic acid. This depresses the pH value of the condensate returns and results in corrosive attack on the equipment and piping.

In general, a dealkalizer is best applied to boilers operating below 700 psi (48 bar). In order to justify installation of a dealkalizer on low-pressure boilers, the alkalinity content should be above 50 ppm with the amount of make-up water exceeding 1,000 gallons (approx. 4,000 litres) per day.

Cooling system make-up will also benefit from reduced alkalinity. The addition of a dealkalizer to a cooling water system will substantially reduce the amount of acid required to treat the same amount of water.

Electric water boiler

An electric water boiler, also called an electric dispensing pot, electric water heater, electric water urn, or electric kettle, is a consumer electronics small appliance used for boiling water and possibly maintaining it at a constant temperature. It is typically used to provide an immediate source of hot water for making tea, hot chocolate, ramen noodles, or baby formula, or any other household use where clean hot water is required.

Components

An electric water boiler consists of a water reservoir with a heating element at the bottom. Some models offer multiple temperature settings. Other models are part of larger water systems that boil water and provide hot, cold, and lukewarm water. Water may be dispensed in various ways, e.g. by pouring, an electric pump or by pressing a large button that functions as a diaphragm pump. Electric water boilers have a built in thermometer that detects when water has reached its boiling point of 100 °C (212 °F) to automatically shut off.

Sedimentation

Sedimentation is the accumulation within the water reservoir of natural minerals that exist in trace amounts in municipal water mains, mainly calcium carbonate. Heating the water causes the minerals to separate and fall to the bottom of the reservoir. This buildup can eventually create a variety of interesting noises in gas boilers, reduce the efficiency of the unit and give rise to a sulfur (or rotten-egg) smell.

Aquastat

An aquastat is a device used in hydronic heating systems for controlling water temperature. To prevent the boiler from firing too often, aquastats have a high limit temperature and a low limit. If the thermostat is calling for heat, the boiler will fire until the high limit is reached, then shut off (even if the thermostat is still calling for heat). The boiler will re-fire if the boiler water temperature drops below a range around the high limit. The high limit exists for the sake of efficiency and safety.

The boiler will also fire (regardless of thermostat state) when the boiler water temperature goes below a range around the low limit, ensuring that the boiler water temperature remains above a certain point. The low limit is intended for tankless domestic hot water---it ensures that boiler water is always warm enough to heat the domestic hot water. Many aquastats also have a "diff" control which determines the size of the range around the "low" and/or "high" controls.

Boiler explosion

A boiler explosion is a catastrophic failure of a boiler. As seen today, boiler explosions are of two kinds. One kind is over-pressure in the pressure parts of the steam and water sides. The second kind is explosion in the furnace. Boiler explosions of pressure parts are particularly associated with steam locomotives. Locomotive boilers are of a construction with a firebox containing the burning fuel, a boiler barrel containing boiling water under pressure, and tubes containing hot gases from the fire (a fire-tube boiler). In these, the latter type of explosion from the furnace side is practically unknown. There can be many different causes, such as failure of the safety valve or corrosion of critical parts of the boiler. Corrosion at the edges of lap joints was a common cause of early boiler explosions.

Principle

It is particularly devastating because of the energy stored up in the heated liquid water. As an example, a basic fire-tube boiler steaming at 50 psi (340 kPa) contains water at a temperature of roughly 150 °C (300 °F). Were a catastrophic failure to occur and the vessel depressurized, most of the water would instantly flash into steam.

Steam takes up 1,600 times more space than liquid water, meaning that each cubic metre of heated boiler water will expand into 1,600 cubic metres of steam and it will do so in a fraction of a second. That enormous volume of steam will expand outwardly to equalize its pressure with the atmosphere and thereby produce an explosion.

In the case of a firebox explosion, these typically occur after a burner flameout. Oil fumes, natural gas, propane, coal, or any other fuel can build up inside the combustion chamber. This is especially of concern when the vessel is hot; the fuels will rapidly volatize due to the temperature. Once the lower explosive limit (LEL) is reached, any source of ignition will cause an explosion of the vapors.

A fuel explosion within the confines of the firebox may damage the pressurized boiler tubes and interior shell, potentially triggering structural failure, steam or water leakage, and/or a secondary boiler shell failure and steam explosion.

Locomotive boilers

These boilers are of a fire-tube type with coke, wood, coal or oil used as fuel. The water feed is by means of steam-powered injectors or boiler feedwater pumps. For storage of fuel and water a separate tender is often provided, adding to the length of the locomotive. This tender usually has a sloping floor for easy flow of fuel toward the locomotive cab and firebox therein. There is a safety valve included on the steam side and also one or more gauges to warn of low water levels.

Any failure of these would result in an explosion of the pressure parts with consequent injury to operating personnel, apart from the damage to equipment. The consequences are more severe due to the restricted working space and constant movement of the locomotives. Safety valves are provided to operate the pressure parts within safe limits. The water level alarms are provided for corrective action by the locomotive crew.

Steamboat boilers

SS Ada Hancock, a small steamboat used to transfer passengers and cargo to and from the large coastal steamships that stopped in San Pedro Harbor in the early 1860s, suffered disaster when its boiler exploded violently in San Pedro Bay, the port of Los Angeles, near Wilmington, California on April 27, 1863 killing twenty-six people and injuring many others of the fifty-three or more passengers on board.

The steamboat Sultana was destroyed in an explosion on 27 April 1865, resulting in the greatest maritime disaster in United States history. An estimated 1,700 passengers were killed when one of the ship's four boilers exploded and the Sultana sank not far from Memphis, Tennessee. The boiler was thought to be the victim of bad construction. Sometimes known as 'the leaky woes'.

Another US Civil War Steamboat explosion was the Steamer Eclipse on January 27, 1865, which was carrying members of the 9th Indiana Artillery. One official Records report mentions the disaster reports 10 killed and 68 injured; a later report mentions that 27 were killed and 78 wounded. Fox's Regimental Losses reports 29 killed.

Use of boilers

The stationary steam engines used to power machinery first came to prominence during the industrial revolution, and in the early days there were many boiler explosions from a variety of causes. One of the first investigators of the problem was William Fairbairn, who helped establish the first insurance company dealing with the losses such explosions could cause. He also established experimentally that the hoop stress in a cylindrical pressure vessel like a boiler was twice the longitudinal stress. Such investigations helped him and others explain the importance of stress concentrations in weakening boilers.

Modern boilers

Modern boilers are designed with redundant pumps, valves, water level monitors, fuel cutoffs, automated controls, and pressure relief valves. In addition, the construction must adhere to strict engineering guidelines set by the relevant authorities. The NBIC, ASME, and others attempt to ensure safe boiler designs by publishing detailed standards. The result is a boiler unit which is less prone to catastrophic accidents.

Also improving safety is the increasing use of "package boilers." These are boilers which are built at a factory then shipped out as a complete unit to the job site. These typically have better quality and fewer problems than boilers which are site assembled tube-by-tube. A package boiler only needs the final connections to be made (electrical, breaching, condensate lines, etc.) to complete the installation.

Explosions

In steam locomotive boilers, as knowledge was gained by trial and error in early days, the explosive situations and consequent damage due to explosions were inevitable. However, improved design and maintenance markedly reduced the number of boiler explosions by the end of the 19th century. Further improvements continued in the 20th century.

On land-based boilers, explosions of the pressure systems happened regularly in stationary steam boilers in the Victorian era, but are now very rare because of the various protections provided, and also because of regular inspections compelled by governmental and industry requirements. Furnace side explosions do happen occasionally, in spite of provisions requiring furnace side explosion doors, wrecking the whole boiler mostly due to operators bypassing the operating instructions.

LOCOMOTIVE BOILER EXPLOSIONS IN THE UK

Hewison (1983) gives a comprehensive account of British boiler explosions, listing 137 between 1815 and 1962. It is noteworthy that 122 of these were in the 19th century and only 15 in the 20th century.

Boiler explosions generally fall into two categories. The first is the breakage of the boiler barrel itself, through weakness/damage or excessive internal pressure, resulting in sudden discharge of steam over a wide area. Boiler plates have been thrown up to a quarter of a mile (Hewison, Rolt). The second type is the collapse of the firebox under steam pressure from the adjoining boiler, releasing flames and hot gases into the cab. Improved design and maintenance almost totally eliminated the first type, but the second type is always possible if the traincrew do not maintain the water level in the boiler.

Boiler barrels could explode if the internal pressure became too high. To prevent this, safety valves were installed to release the pressure at a set level. Early examples were spring-loaded, but John Ramsbottom invented a tamper-proof valve which was universally adopted. The other common cause of explosions was internal corrosion which weakened the boiler barrel so that it could not withstand normal operating pressure. In particular, grooves could occur along horizontal seams (lap joints) below water level. Dozens of explosions resulted, but were eliminated by 1900 by the adoption of butt joints, plus improved maintenance schedules and regular hydraulic testing.

Fireboxes were generally made of copper, though later locomotives had steel fireboxes. They were held to the outer part of the boiler by stays (numerous small supports). Parts of the firebox in contact with full steam pressure have to be kept covered with water, to stop them overheating and weakening.

The usual cause of firebox collapses is that the boiler water level falls too low and the top of the firebox (crown sheet) becomes uncovered and overheats. This occurs if the fireman has failed to maintain water level or the level indicator (gauge glass) is faulty. A less common reason is breakage of large numbers of stays, due to corrosion or unsuitable material.

Throughout the 20th century, two boiler barrel failures and thirteen firebox collapses occurred. The boiler barrel failures occurred at Cardiff in 1909 and Buxton in 1921; both were caused by misassembly of the safety valves causing the boilers to exceed their design pressures. Of the 13 firebox collapses, four were due to broken stays, one to scale buildup on the firebox, and the rest were due to low water level.

Relief Valve

The relief valve (RV) is a type of valve used to control or limit the pressure in a system or vessel which can build up by a process upset, instrument or equipment failure, or fire.

The pressure is relieved by allowing the pressurised fluid to flow from an auxiliary passage out of the system. The relief valve is designed or set to open at a predetermined set pressure to protect pressure vessels and other equipment from being subjected to pressures that exceed their design limits.

When the set pressure is exceeded, the relief valve becomes the "path of least resistance" as the valve is forced open and a portion of the fluid is diverted through the auxiliary route. The diverted fluid (liquid, gas or liquid-gas mixture) is usually routed through a piping system known as a flare header or relief header to a central, elevated gas flare where it is usually burned and the resulting combustion gases are released to the atmosphere.

As the fluid is diverted, the pressure inside the vessel will drop. Once it reaches the valve's reseating pressure, the valve will close. The blowdown is usually stated as a percentage of set pressure and refers to how much the pressure needs to drop before the valve reseats. The blowdown can vary from roughly 2-20%, and some valves have adjustable blowdowns.

In high-pressure gas systems, it is recommended that the outlet of the relief valve is in the open air. In systems where the outlet is connected to piping, the opening of a relief valve will give a pressure build up in the piping system downstream of the relief valve. This often means that the relief valve will not re-seat once the set pressure is reached. For these systems often so called "differential" relief valves are used.

This means that the pressure is only working on an area that is much smaller than the openings area of the valve. If the valve is opened the pressure has to decrease enormously before the valve closes and also the outlet pressure of the valve can easily keep the valve open. Another consideration is that if other relief valves are connected to the outlet pipe system, they may open as the pressure in exhaust pipe system increases. This may cause undesired operation.

In some cases, a so-called bypass valve acts as a relief valve by being used to return all or part of the fluid discharged by a pump or gas compressor back to either a storage reservoir or the inlet of the pump or gas compressor. This is done to protect the pump or gas compressor and any associated equipment from excessive pressure. The bypass valve and bypass path can be internal (an integral part of the pump or compressor) or external (installed as a component in the fluid path). Many fire engines have such relief valves to prevent the overpressurization of fire hoses.

In other cases, equipment must be protected against being subjected to an internal vacuum (i.e., low pressure) that is lower than the equipment can withstand. In such cases, vacuum relief valves are used to open at a predetermined low pressure limit and to admit air or an inert gas into the equipment so as control the amount of vacuum.

Technical Terms

In the petroleum refining, petrochemical and chemical manufacturing, natural gas processing and power generation industries, the term relief valve is associated with the terms pressure relief valve (PRV), pressure safety valve (PSV) and safety valve.

In practice, people often do not stick to the technical distinctions between the most common abbreviations: SRV, PRV, SV and RV

Pressure Relief Valve (PRV) or Pressure Safety Valve (PSV). The difference being that PSVs have a manual lever to activate the valve in case of emergency. Most PRV are spring operated. At lower pressures some use a diaphragm in place of a spring. The oldest PRV designs use a weight to seal the valve.

Set Pressure: When increasing system pressure reaches this value the PRV opens. Accuracy of set pressure often follows guidelines set by the ASME. Relief valve (RV): A valve used on a liquid service, which opens proportionally as the increasing pressure overcomes the spring pressure.

Safety valve (SV): Used in gas service. Most SV are full lift or snap acting, they pop open all the way.

Safety relief valve (SRV): A PRV that can be used for gas or liquid service. But set pressure will usually only be accurate for one type of fluid at a time (the type it was set with).

Pilot-operated relief valve (POSRV, PORV, POPRV): device that relieves by remote command from a pilot valve that is connected to the upstream system pressure. Low pressure safety valve (LPSV): automatic system that relieves by static pressure on a gas. The pressure is small and near the atmospheric pressure.

Vacuum pressure safety valve (VPSV): automatic system that relieves by static pressure on a gas. The pressure is small, negative and near the atmospheric pressure.

Low and vacuum pressure safety valve (LVPSV): automatic system that relieves by static pressure on a gas. The pressure is small, negative or positive and near the atmospheric pressure.

Snap Acting: The opposite of modulating, refers to a valve that "pops" open, it goes into full lift in milliseconds. Usually accomplished with a skirt on the disc so that the fluid passing the seat suddenly affects a larger area and creates more lifting force.

Modulating: Opens in proportion to the overpressure formed.

Legal and code requirements in industry

In most countries, industries are legally required to protect pressure vessels and other equipment by using relief valves. Also in most countries, equipment design codes such as those provided by the American Society of Mechanical Engineers (ASME), American Petroleum Institute (API) and other organizations like ISO (ISO 4126) must be complied with and those codes include design standards for relief valves.

The main standards, laws or directives are:

- ASME (American Society of Mechanical Engineers) Boiler & Pressure Vessel Code, Section VIII Division 1 and Section I
- API (American Petroleum Institute) Recommended Practice 520/521, API Standard 2000 et API Standard 526
- ISO 4126 (International Organisation for Standardisation)
- EN 764-7 (European Standard based on pressure Equipment Directive 97/23/EC)
- AD Merkblatt (German)
- PED 97/23/EC (Pressure Equipment Directive - European Union)

DIERS

Formed in 1976, the Design Institute for Emergency Relief Systems was a consortium of 29 companies under the auspices of the American Institute of Chemical Engineers (AIChE) that developed methods for the design of emergency relief systems to handle runaway reactions. Its purpose was to develop the technology and methods needed for sizing pressure relief systems for chemical reactors, particularly those in which exothermic reactions are carried out.

Such reactions include many classes of industrially important processes including polymerizations, nitrations, diazotizations, sulphonations, epoxidations, aminations, esterifications, neutralizations and many others. Pressure relief systems can be difficult to design, not least because what is expelled can be gas/vapour, liquid, or a mixture of the two - just as with a can of carbonated drink when it is suddenly opened. For chemical reactions, it requires extensive knowledge of both chemical reaction hazards and fluid flow.

DIERS investigated the two-phase vapor-liquid onset / disengagement dynamics and the hydrodynamics of emergency relief systems with extensive experimental and analysis work. Of particular interest to DIERS were the prediction of two-phase flow venting and the applicability of various sizing methods for two-phase vapor-liquid flashing flow. DIERS became a user's group in 1985.

European DIERS Users' Group (EDUG) is a group of mainly European industrialists, consultants and academics who use the DIERS technology. The EDUG started in the late 1980s and has an annual meeting. A summary of many of key aspects of the DIERS technology has been published in the UK by the HSE.

Safety Valve

A safety valve is a valve mechanism for the automatic release of a substance from a boiler, pressure vessel, or other system when the pressure or temperature exceeds preset limits. It is part of a bigger set named pressure safety valves (PSV) or pressure relief valves (PRV). The other parts of the set are named relief valves, safety relief valves, pilot-operated relief valves, low pressure safety valves, vacuum pressure safety valves.

Safety valves were first used on steam boilers during the industrial revolution. Early boilers without them were prone to accidental explosion. Vacuum safety valves (or combined pressure / vacuum safety valves) are used to prevent a tank to collapse when emptying it or when cold rinse water is used after hot CIP or SIP. The calculation method is not defined in any norm when sizing a vacuum safety valve, particularly in the hot CIP / cold water scenario, but some manufacturers have developed simulations to do so.

Function and design

The earliest and simplest safety valve on the steam digester in 1679 used a weight to hold the pressure of the steam, (this design is still commonly used on pressure cookers); however, these were easily tampered with or accidentally released. On the Stockton and Darlington Railway, the safety valve tended to go off when the engine hit a bump in the track.

A valve less sensitive to sudden accelerations used a spring to contain the steam pressure, but these (based on Salter spring balances) could still be screwed down to increase the pressure beyond design limits. This dangerous practice was sometimes used to marginally increase performance of a steam engine. In 1856, John Ramsbottom invented a tamper-proof spring safety valve which became universal on railways.

Safety valves also evolved to protect equipment such as pressure vessels (fired or not) and heat exchangers. The term safety valve should be limited to compressible fluid application (gas, vapor, steam).

The two general types of protection encountered in industry are thermal protection and flow protection.

For liquid-packed vessels, thermal relief valves are generally characterized by the relatively small size of the valve necessary to provide protection from excess pressure caused by thermal expansion. In this case a small valve is adequate because most liquids are nearly incompressible, and so a relatively small amount of fluid discharged through the relief valve will produce a substantial reduction in pressure.

Flow protection is characterized by safety valves that are considerably larger than those mounted in thermal protection. They are generally sized for use in situations where significant quantities of gas or high volumes of liquid must be quickly discharged in order to protect the integrity of the vessel or pipeline. This protection can alternatively be achieved by installing a high integrity pressure protection system (HIPPS).

Technical terms

In the petroleum refining, petrochemical, chemical manufacturing, natural gas processing, power generation, food, drinks, cosmetics ans pharmaceuticals industries, the term safety valve is associated with the terms pressure relief valve (PRV), pressure safety valve (PSV) and relief valve. The generic term is Pressure relief valve (PRV) or pressure safety valve (PSV) It should be noted, as most people think PRV and PSV are the same thing. Is that PSV's have a manual lever to open the valve in case of emergency.

- Relief valve (RV): automatic system that is actuated by static pressure in a liquid-filled vessel. It specifically opens proportionally with increasing pressure.
- Safety valve (SV): automatic system that relieves the static pressure on a gas. It usually opens completely, accompanied by a popping sound.
- Safety relief valve (SRV): automatic system that relieves by static pressure on both gas and liquid.
- Pilot-operated safety relief valve (POSRV): automatic system that relieves by remote command from a pilot on which the static pressure (from equipment to protect) is connected.
- Low pressure safety valve (LPSV): automatic system that relieves static pressure on a gas. Used when difference between vessel pressure and the ambient atmospheric pressure is small.
- Vacuum pressure safety valve (VPSV): automatic system that relieves static pressure on a gas. Used when the pressure difference between the vessel pressure and the ambient pressure is small, negative and near the atmospheric pressure.
- Low and vacuum pressure safety valve (LVPSV): automatic system that relieves static pressure on a gas. The pressure is small, negative or positive and near the atmospheric pressure.

RV, SV and SRV are spring operated (even said spring loaded). LPSV and VPSV are spring operated or weight loaded.

LEGAL AND CODE REQUIREMENTS IN INDUSTRY

In most countries, industries are legally required to protect pressure vessels and other equipment by using relief valves. Also in most countries, equipment design codes such as those provided by the ASME, API and other organizations like ISO (ISO 4126) must be complied with and those codes include design standards for relief valves. Today, industries such as food, drinks, cosmetics, pharmaceuticals and fine chemicals industries ask for hygienic safety valves, fully drainable and Cleanable In Place; most of them are stainless steel made; Hygienic norms are mainly 3A in the US and EHEDG in Europe.

Types

There is a wide range of safety valves having many different applications and performance criteria required to cover different areas. In addition, national standards are set by many kinds of safety valve.

- ASME I tap - a safety valve in accordance with the requirements of Section I of the application code ASME pressure vessel, which opens 3% and 4% of the pressure. Will rule on two rings serving and supported by a National Seal V defined.
- ASME VIII valve - safety valve in accordance with the requirements of Article VIII of the ASME code for pressure vessel applications, which is within 10% overpressure that opens and closes in 7%. Characterized by a National Board UV stamp.
- Low-lift safety valve - the current position of the disc around the drain valve.
- Full lift safety valve - the region of the exemption is not determined by the position of the disc.
- Full-flow relief valve - A valve which is expected in the hole and lift the valve a sufficient measure of the minimum area for each position or under the seat to make the control panel.

Safety Valve * classic - The spring housing is vented to the pressure side, i.e., functional characteristics are directly influenced by changes in pressure in the valve.

- Valve balanced - a balanced valve includes a means to minimize the effects of pressure on the operating characteristics of the valve.

- Guide pressure valve - The largest dump device is in conjunction with controlled, self-actuated auxiliary pressure relief.
- Power operated valve - A pressure relief valve in which the main system to relieve pressure combined with and controlled by a device requires an external power source.
- Is the standard safety valve - a valve opening to achieve the degree of buoyancy required for mass flow increases the pressure to reject over 10%. (The valve is characterized by a score of popular art, sometimes also called high-lift).
- Full-lift (solid line) valve - a valve that opens at the beginning of the elevator, quickly, with a 5% increase in pressure until complete removal is limited by design. The amount of lift to an early start (analog) can not exceed 20%.
- Valve and spring - a safety valve for the treatment of violent opening of the valve plate by a clamping force as a spring or weight.
- Proportional-relief valve - a safety valve which opens more or less stable in comparison with increasing pressure. sudden opening within a stroke by 10% will not happen without increasing pressure. After opening at a pressure greater than 10% meet these safety valves of the lift and landed on the mass flowmeter.
- Safety valve diaphragm - A direct-loaded safety valve with linear and rotary components and springs to protect against the effects of the liquid membrane.
- Bellows valve - A direct-loaded safety valve and slide (or part) of rotating elements and sources are protected from the effects of fluid through bellows. The bellows allows such an interpretation, it is as if the influences of pressure.
- Controlled valve - A valve that consists of a main valve and a control unit. It also includes direct acting safety valves with additional load, which, until the total pressure reached, an additional force increases the closing force.
- Safety valve - A valve that automatically, without using any form of energy than the liquid in the discharge quantity of liquid, so as to prevent a predetermined safe pressure is higher, and closed again to prevent the flow of more volatile after the normal working pressure were restored. Note that the valve can be characterized by a pop-action (quick opening) or open relationship (not necessarily linear) on increasing pressure over the whole pressure.

- Valve and spring - a safety valve when the charger through the liquid under pressure valve plate with mechanical pressure directly, unlike, for example, a weight, lever and weight or spring. * Assisted valve - A valve which can be lifted by a support mechanism for moving, also gradually the pressure that the pressure setting, and for failure to comply with a mechanism to support all requirements for safety valves in the standard.
- Safety valve spring in addition - as a safety valve until the inlet pressure safety valve to the pressure achieved additional power increases the power of foreclosure. Note that this extra power (extra weight), which is made from an external power source is available, reliable published when the inlet pressure reaches the safety valve pressure set. The amount of the extra load is placed so that if such a charge is released, the safety valve of certified capacity at a pressure not greater than 1. protected 1 times the maximum allowable pressure devices.
- Master valve - A valve whose operation is initiated and controlled by the fluid discharged from a pilot valve is a safety net direct costs, provided that standard.

United States

- ASME (American Society of Mechanical Engineers) Boiler & Pressure Vessel Code, Section I
- ASME (American Society of Mechanical Engineers) Boiler & Pressure Vessel Code, Section VIII, Division 1
- API (American Petroleum Institute) Recommended Practice 520 and API Standard 526, API Standard 2000 (low pressure - Storage tank)

European Union

European standard steam boiler safety valve

- ISO 4126 (harmonized with European Union directives)
- EN 764-7 (former CEN standard, harmonized with European Union directives, replaced with EN ISO 4126-1)
- AD Merkblatt (German)
- PED 97/23/EC (Pressure Equipment Directive - European Union)

Water heaters

They are required on water heaters, where they prevent disaster in certain configurations in the event a thermostat should fail. There are still occasional, spectacular failures of older water heaters that lack this equipment. Houses can be leveled by the force of the blast.

Pressure cookers

Pressure cookers are pots for cooking with a pressure-proof lid. Cooking at pressure allows the temperature to rise above the normal boiling point of water (100 degrees Celsius at sea level) which speeds up the cooking and makes the cooking more thorough.

Pressure cookers usually have two safety valves. One is a hole upon which a weight sits. The other is a sealed rubber grommet which is ejected in a controlled explosion if the first valve gets blocked.

7

ENGINES

Originally, an engine was a mechanical device that converted force into motion. Military devices such as catapults, trebuchets and battering rams are referred to as siege engines. Most devices used in the industrial revolution were referred to as engines, and this is where the steam engine gained its name. In modern usage, the term is used to describe devices capable of performing mechanical work, as in the original steam engine.

In most cases, the work is produced by exerting a torque or linear force, which is used to operate other machinery which can generate electricity, pump water, or compress gas. In the context of propulsion systems, an air-breathing engine is one that uses atmospheric air to oxidise the fuel carried rather than supplying an independent oxidizer, as in a rocket.

In common usage, an engine burns or otherwise consumes fuel, and is differentiated from an electric machine (i.e., electric motor) that derives power without changing the composition of matter. A heat engine may also serve as a prime mover, a component that transforms the flow or changes in pressure of a fluid into mechanical energy. An automobile powered by an internal combustion engine may make use of various motors and pumps, but ultimately all such devices derive their power from the engine.

The term motor was originally used to distinguish the new internal combustion engine-powered vehicles from earlier vehicles powered by steam engines, such as the steam roller and motor roller, but may be used to refer to any engine

An engine or motor is a machine designed to convert energy into useful mechanical motion. Motors converting heat energy into motion are usually referred to as engines, which come in many types. A common type is a heat engine such as an internal combustion engine which typically burns a fuel with air and uses the hot gases for generating power. External combustion engines such as steam engines use heat to generate motion via a separate working fluid.

Another common type of motor is the electric motor. This takes electrical energy and generates mechanical motion via varying electromagnetic fields. Other motors including pneumatic motors that are driven by compressed air, and motors can be driven by elastic energy, such as springs. Some motors are driven by non combustive chemical reactions.

HISTORY OF ENGINES

Antiquity

Simple machines, such as the club and oar (examples of the lever), are prehistoric. More complex engines using human power, animal power, water power, wind power and even steam power date back to antiquity. Human power was focused by the use of simple engines, such as the capstan, windlass or treadmill, and with ropes, pulleys, and block and tackle arrangements; this power was transmitted usually with the forces multiplied and the speed reduced. These were used in cranes and aboard ships in Ancient Greece, as well as in mines, water pumps and siege engines in Ancient Rome. The writers of those times, including Vitruvius, Frontinus and Pliny the Elder, treat these engines as commonplace, so their invention may be far more ancient. By the 1st century AD, various breeds of cattle and horses were used in mills, driving machines similar to those powered by humans in earlier times.

According to Strabo, a water powered mill was built in Kaberia of the kingdom of Mithridates during the 1st century BC. Use of water wheels in mills spread throughout the Roman Empire over the next few centuries. Some were quite complex, with aqueducts, dams, and sluices to maintain and channel the water, along with systems of gears, or toothed-wheels made of wood and metal to regulate the speed of rotation. In a poem by Ausonius in the 4th century AD, he mentions a stone-cutting saw powered by water. Hero of Alexandria is credited with many such wind and steam powered machines in the 1st century AD, including the Aeolipile, but it is not known if any of these were put to practical use.

Medieval

This section may contain inappropriate or misinterpreted citations that do not verify the text. Please help improve this article by checking for inaccuracies. (help, talk, get involved!) (September 2010)

During the Muslim Agricultural Revolution from the 9th to 13th centuries, Muslim engineers developed numerous innovative industrial uses of hydropower, early industrial uses of tidal power, wind power, and fossil fuels such as petroleum, together with the earliest large factory complexes (tiraz in Arabic). The industrial uses of watermills in the Islamic world date back to the 7th century, whereas horizontal-wheeled and vertical-wheeled water mills were both in widespread use since at least the 9th century. A variety of industrial mills were invented in the Islamic world, including fulling mills, hullers, steel mills, sugar refineries, and windmills. By the 11th century, every province throughout the Islamic world had these industrial mills in operation, from the Middle East and Central Asia to al-Andalus and North Africa.

Roman engineers invented water turbines in the 4th century AD, Muslim engineers employed gears in mills and water-raising machines, and pioneered the use of dams as a source of water power to provide additional power to watermills and water-raising machines. Such advances made it possible for many industrial tasks that were previously driven by manual labour to be mechanized and driven by machinery to some extent in the medieval Islamic world.

In 1206, al-Jazari employed a crank-connecting rod system for two of his water-raising machines. A similar steam turbine later appeared in Europe a century later, which eventually led to the steam engine and Industrial Revolution in 18th century Europe.

Industrial revolution

English inventor Sir Samuel Morland allegedly used gunpowder to drive water pumps in the 17th century. For more conventional, reciprocating internal combustion engines, the fundamental theory for two-stroke engines was established by Sadi Carnot, France, 1824, whilst the American Samuel Morey received a patent on April 1, 1826. Sir Dugald Clark (1854-1932) designed the first two-stroke engine in 1878 and patented it in England in 1881. Automotive engines have used a range of energy-conversion systems. These include electric, steam, solar, turbine, rotary, piston-type internal combustion engine.

Karl Benz was one of the leaders in the development of new engines. In 1878 he began to work on new designs. He concentrated his efforts on creating a reliable gas two-stroke engine that was more powerful, based on Nikolaus Otto's design of the four-stroke engine. Karl Benz showed his real genius, however, through his successive inventions registered while designing what would become the production standard for his two-stroke engine. Benz was granted a patent for it in 1879.

The lightweight petrol internal combustion engine, operating on a four-stroke Otto cycle, has been the most successful for automobiles, while the more efficient diesel engine is used for trucks and buses.

HORIZONTALLY OPPOSED PISTONS

In 1896, Karl Benz was granted a patent for his design of the first engine with horizontally opposed pistons. Many BMW motorcycles use this engine type. His design created an engine in which the corresponding pistons move in horizontal cylinders and reach top dead center simultaneously, thus automatically balancing each other with respect to their individual momentums. Engines of this design are often referred to as flat engines because of their shape and lower profile. They must have an even number of cylinders and six, four or two cylinder flat engines have all been common. The most well-known engine of this type is probably the Volkswagen Beetle engine. Engines of this type continue to be a common design principle for high performance aero engines (for propeller driven aircraft) and, engines used by automobile producers such as Porsche and Subaru.

Advancement

Continuance of the use of the internal combustion engine for automobiles is partly due to the improvement of engine control systems (onboard computers providing engine management processes, and electronically controlled fuel injection). Forced air induction by turbocharging and supercharging have increased power outputs and engine efficiencies.

Similar changes have been applied to smaller diesel engines giving them almost the same power characteristics as petrol engines. This is especially evident with the popularity of smaller diesel engine propelled cars in Europe. Larger diesel engines are still often used in trucks and heavy machinery.

They do not burn as clean as gasoline engines, however they have far more torque. The internal combustion engine was originally selected for the automobile due to its flexibility over a wide range of speeds. Also, the power developed for a given weight engine was reasonable; it could be produced by economical mass-production methods; and it used a readily available, moderately priced fuel - petrol.

Increasing power

The first half of the 20th century saw a trend to increasing engine power, particularly in the American models. Design changes incorporated all known methods of raising engine capacity, including increasing the pressure in the cylinders to improve efficiency, increasing the size of the engine, and increasing the speed at which power is generated. The higher forces and pressures created by these changes created engine vibration and size problems that led to stiffer, more compact engines with V and opposed cylinder layouts replacing longer straight-line arrangements.

Combustion efficiency

The design principles favoured in Europe, because of economic and other restraints such as smaller and twistier roads, leant toward smaller cars and corresponding to the design principles that concentrated on increasing the combustion efficiency of smaller engines. This produced more economical engines with earlier four-cylinder designs rated at 40 horsepower (30 kW) and six-cylinder designs rated as low as 80 horsepower (60 kW), compared with the large volume V-8 American engines with power ratings in the range from 250 to 350 hp (190 to 260 kW).

Engine configuration

Earlier automobile engine development produced a much larger range of engines than is in common use today. Engines have ranged from 1 to 16 cylinder designs with corresponding differences in overall size, weight, piston displacement, and cylinder bores. Four cylinders and power ratings from 19 to 120 hp (14 to 90 kW) were followed in a majority of the models. Several three-cylinder, two-stroke-cycle models were built while most engines had straight or in-line cylinders. There were several V-type models and horizontally opposed two- and four-cylinder makes too. Overhead camshafts were frequently employed.

The smaller engines were commonly air-cooled and located at the rear of the vehicle; compression ratios were relatively low. The 1970s and '80s saw an increased interest in improved fuel economy which brought in a return to smaller V-6 and four-cylinder layouts, with as many as five valves per cylinder to improve efficiency. The Bugatti Veyron 16.4 operates with a W16 engine meaning that two V8 cylinder layouts are positioned next to each other to create the W shape.

The largest internal combustion engine ever built is the Wärtsilä-Sulzer RTA96-C, a 14-cylinder, 2-stroke turbocharged diesel engine that was designed to power the Emma Maersk, the largest container ship in the world. This engine weighs 2300 tons, and when running at 102 RPM produces 109,000 bhp (80,080 kW) consuming some 13.7 tons of fuel each hour.

Heat Engine

In thermodynamics, heat engines are often modeled using a standard engineering model such as the Otto cycle. The theoretical model can be refined and augmented with actual data from an operating engine, using tools such as an indicator diagram. Since very few actual implementations of heat engines exactly match their underlying thermodynamic cycles, one could say that a thermodynamic cycle is an ideal case of a mechanical engine.

In any case, fully understanding an engine and its efficiency requires gaining a good understanding of the (possibly simplified or idealized) theoretical model, the practical nuances of an actual mechanical engine, and the discrepancies between the two.

In general terms, the larger the difference in temperature between the hot source and the cold sink, the larger is the potential thermal efficiency of the cycle. On Earth, the cold side of any heat engine is limited to being close to the ambient temperature of the environment, or not much lower than 300 Kelvin, so most efforts to improve the thermodynamic efficiencies of various heat engines focus on increasing the temperature of the source, within material limits. The maximum theoretical efficiency of a heat engine (which no engine ever obtains) is equal to the temperature difference between the hot and cold ends divided by the temperature at the hot end, all expressed in absolute temperature or kelvins.

The efficiency of various heat engines proposed or used today ranges from 3 percent (97 percent waste heat) for the OTEC ocean power proposal through 25 percent for most automotive engines, to 45 percent for a supercritical coal plant, to about 60 percent for a steam-cooled combined cycle gas turbine. All of these processes gain their efficiency (or lack thereof) due to the temperature drop across them.

Power

Heat engines can be characterized by their specific power, which is typically given in kilowatts per litre of engine displacement (in the U.S. also horsepower per cubic inch). The result offers an approximation of the peak-power output of an engine. This is not to be confused with fuel efficiency, since high-efficiency often requires a lean fuel-air ratio, and thus lower power density. A modern high-performance car engine makes in excess of 75 kW/L (1.65 hp/in^3).

Everyday examples

Examples of everyday heat engines include the steam engine, the diesel engine, and the gasoline (petrol) engine in an automobile. A common toy that is also a heat engine is a drinking bird. Also the stirling engine is a heat engine. All of these familiar heat engines are powered by the expansion of heated gases. The general surroundings are the heat sink, providing relatively cool gases which, when heated, expand rapidly to drive the mechanical motion of the engine.

Examples of heat engines

It is important to note that although some cycles have a typical combustion location (internal or external), they often can be implemented as the other combustion cycle.

For example, John Ericsson developed an external heated engine running on a cycle very much like the earlier Diesel cycle. In addition, the externally heated engines can often be implemented in open or closed cycles. What this boils down to is there are thermodynamic cycles and a large number of ways of implementing them with mechanical devices called engines.

Phase change cycles

In these cycles and engines, the working fluids are gases and liquids. The engine converts the working fluid from a gas to a liquid, from liquid to gas, or both, generating work from the fluid expansion or compression.

- Rankine cycle (classical steam engine)
- Regenerative cycle (steam engine more efficient than Rankine cycle)
- Organic Rankine Cycle (Coolant changing phase in temperature ranges of ice and hot liquid water)
- Vapor to liquid cycle (Drinking bird, Injector, Minto wheel)
- Liquid to solid cycle (Frost heaving - water changing from ice to liquid and back again can lift rock up to 60 cm.)
- Solid to gas cycle (Dry ice cannon - Dry ice sublimes to gas.)

Gas only cycles

In these cycles and engines the working fluid is always a gas (i.e., there is no phase change):

- Carnot cycle (Carnot heat engine)
- Ericsson Cycle (Caloric Ship John Ericsson)
- Stirling cycle (Stirling engine, thermoacoustic devices)
- Internal combustion engine (ICE):
  - Otto cycle (e.g. Gasoline/Petrol engine, high-speed diesel engine)
  - Diesel cycle (e.g. low-speed diesel engine)
  - Atkinson Cycle (Atkinson Engine)
  - Brayton cycle or Joule cycle originally Ericsson Cycle (gas turbine)

- o Lenoir cycle (e.g., pulse jet engine)
- o Miller cycle

Liquid only cycles

In these cycles and engines the working fluid are always like liquid:

- Stirling Cycle (Malone engine)
- Heat Regenerative Cyclone

Electron cycles

- Johnson thermoelectric energy converter
- Thermoelectric (Peltier-Seebeck effect)
- Thermionic emission
- Thermotunnel cooling

Magnetic cycles

- Thermo-magnetic motor (Tesla)

Cycles used for refrigeration

A domestic refrigerator is an example of a heat pump: a heat engine in reverse. Work is used to create a heat differential. Many cycles can run in reverse to move heat from the cold side to the hot side, making the cold side cooler and the hot side hotter. Internal combustion engine versions of these cycles are, by their nature, not reversible.

Refrigeration cycles include:

- Vapor-compression refrigeration
- Stirling cryocoolers
- Gas-absorption refrigerator
- Air cycle machine
- Vuilleumier refrigeration
- Magnetic refrigeration

Evaporative Heat Engines

The Barton evaporation engine is a heat engine based on a cycle producing power and cooled moist air from the evaporation of water into hot dry air.

Efficiency

The efficiency of a heat engine relates how much useful work is output for a given amount of heat energy input.

From the laws of thermodynamics:

where

dW = – PdV is the work extracted from the engine. (It is negative since work is done by the engine.)

dQh = ThdSh is the heat energy taken from the high temperature system. (It is negative since heat is extracted from the source, hence (– dQh) is positive.)

dQc = TcdSc is the heat energy delivered to the cold temperature system. (It is positive since heat is added to the sink.)

In other words, a heat engine absorbs heat energy from the high temperature heat source, converting part of it to useful work and delivering the rest to the cold temperature heat sink.

In general, the efficiency of a given heat transfer process (whether it be a refrigerator, a heat pump or an engine) is defined informally by the ratio of "what you get out" to "what you put in."

In the case of an engine, one desires to extract work and puts in a heat transfer.

$$\eta = \frac{-dW}{-dQ_h} = \frac{-dQ_h - dQ_c}{-dQ_h} = 1 - \frac{dQ_c}{-dQ_h}$$

The theoretical maximum efficiency of any heat engine depends only on the temperatures it operates between. This efficiency is usually derived using an ideal imaginary heat engine such as the Carnot heat engine, although other engines using different cycles can also attain maximum efficiency. Mathematically, this is because in reversible processes, the change in entropy of the cold reservoir is the negative of that of the hot reservoir (i.e., dSc = – dSh), keeping the overall change of entropy zero. Thus:

$$\eta_{\text{max}} = 1 - \frac{T_c dS_c}{-T_h dS_h} = 1 - \frac{T_c}{T_h}$$

where Th is the absolute temperature of the hot source and Tc that of the cold sink, usually measured in kelvin. Note that dSc is positive while dSh is negative; in any reversible work-extracting process, entropy is overall not increased, but rather is moved from a hot (high-entropy) system to a cold (low-entropy one), decreasing the entropy of the heat source and increasing that of the heat sink.

The reasoning behind this being the maximal efficiency goes as follows. It is first assumed that if a more efficient heat engine than a Carnot engine is possible, then it could be driven in reverse as a heat pump. Mathematical analysis can be used to show that this assumed combination would result in a net decrease in entropy. Since, by the second law of thermodynamics, this is statistically improbable to the point of exclusion, the Carnot efficiency is a theoretical upper bound on the reliable efficiency of any process.

Empirically, no engine has ever been shown to run at a greater efficiency than a Carnot cycle heat engine.

Endoreversible heat engines

The most Carnot efficiency as a criterion of heat engine performance is the fact that by its nature, any maximally efficient Carnot cycle must operate at an infinitesimal temperature gradient. This is because any transfer of heat between two bodies at differing temperatures is irreversible, and therefore the Carnot efficiency expression only applies in the infinitesimal limit. The major problem with that is that the object of most heat engines is to output some sort of power, and infinitesimal power is usually not what is being sought.

A different measure of ideal heat engine efficiency is given by considerations of endoreversible thermodynamics, where the cycle is identical to the Carnot cycle except in that the two processes of heat transfer are not reversible (Callen 1985):

$$\eta = 1 - \sqrt{\frac{T_c}{T_h}}$$

(Note: Units K or °R)

This model does a better job of predicting how well real-world heat engines can do (Callen 1985, see also endoreversible thermodynamics):

As shown, the endoreversible efficiency much more closely models the observed data.

HEAT ENGINE ENHANCEMENTS

Engineers have studied the various heat engine cycles extensively in an effort to improve the amount of usable work they could extract from a given power source. The Carnot Cycle limit cannot be reached with any gas-based cycle, but engineers have worked out at least two ways to possibly go around that limit, and one way to get better efficiency without bending any rules.

1. Increase the temperature difference in the heat engine. The simplest way to do this is to increase the hot side temperature, which is the approach used in modern combined-cycle gas turbines. Unfortunately, physical limits (such as the melting point of the materials from which the engine is constructed) and environmental concerns regarding NOx production restrict the maximum temperature on workable heat engines. Modern gas turbines run at temperatures as high as possible within the range of temperatures necessary to maintain acceptable NOx output. Another way of increasing efficiency is to lower the output temperature. One new method of doing so is to use mixed chemical working fluids, and then exploit the changing behavior of the mixtures. One of the most famous is the so-called Kalina cycle, which uses a 70/30 mix of ammonia and water as its working fluid. This mixture allows the cycle to generate useful power at considerably lower temperatures than most other processes.
2. Exploit the physical properties of the working fluid. The most common such exploitation is the use of water above the so-called critical point, or so-called supercritical steam. The behavior of fluids above their critical point changes radically, and with materials such as water and carbon dioxide it is possible to exploit those changes in behavior to extract greater thermodynamic efficiency from the heat engine, even if it is using a fairly conventional Brayton or Rankine cycle. A newer and very promising material for such applications is CO2. SO2 and xenon have also been considered for such applications, although SO2 is a little toxic for most.
3. Exploit the chemical properties of the working fluid. A fairly new and novel exploit is to use exotic working fluids with advantageous chemical properties. One such is nitrogen dioxide (NO2), a toxic component of smog, which has a natural dimer as di-nitrogen tetraoxide (N2O4). At low temperature, the N2O4 is compressed and then heated. The increasing temperature causes each N2O4 to break apart into two NO2 molecules. This lowers the molecular weight of the working fluid, which drastically increases the efficiency of the cycle. Once the NO2 has expanded through the turbine, it is cooled by the heat sink, which causes it to recombine into N2O4. This is then fed back to the compressor for another cycle. Such species as aluminium bromide (Al2Br6), NOCl, and Ga2I6 have all been investigated for such uses. To date, their drawbacks have not warranted their use, despite the efficiency gains that can be realized.

Each process is one of the following:

- Isothermal (at constant temperature, maintained with heat added or removed from a heat source or sink)
- Isobaric (at constant pressure)
- Isometric/isochoric (at constant volume), also referred to as iso-volumetric
- Adiabatic (no heat is added or removed from the system during adiabatic process which is equivalent to saying that the entropy remains constant)

Internal combustion engine

The internal combustion engine is an engine in which the combustion of a fuel (normally a fossil fuel) occurs with an oxidizer (usually air) in a combustion chamber. In an internal combustion engine, the expansion of the high-temperature and -pressure gases produced by combustion applies direct force to some component of the engine, such as pistons, turbine blades, or a nozzle. This force moves the component over a distance, generating useful mechanical energy.

The term internal combustion engine usually refers to an engine in which combustion is intermittent, such as the more familiar four-stroke and two-stroke piston engines, along with variants, such as the six-stroke piston engine and the Wankel rotary engine. A second class of internal combustion engines use continuous combustion: gas turbines, jet engines and most rocket engines, each of which are internal combustion engines on the same principle as previously described.

The internal combustion engine (or ICE) is quite different from external combustion engines, such as steam or Stirling engines, in which the energy is delivered to a working fluid not consisting of, mixed with, or contaminated by combustion products. Working fluids can be air, hot water, pressurized water or even liquid sodium, heated in some kind of boiler.

A large number of different designs for ICEs have been developed and built, with a variety of different strengths and weaknesses. Powered by an energy-dense fuel (which is very frequently gasoline, a liquid derived from fossil fuels). While there have been and still are many stationary applications, the real strength of internal combustion engines is in mobile applications and they dominate as a power supply for cars, aircraft, and boats, from the smallest to the largest.

An automobile engine partly opened and colored to show components.

Applications

Internal combustion engines are most commonly used for mobile propulsion in vehicles and portable machinery. In mobile equipment, internal combustion is advantageous since it can provide high power-to-weight ratios together with excellent fuel energy density. Generally using fossil fuel (mainly petroleum), these engines have appeared in transport in almost all vehicles (automobiles, trucks, motorcycles, boats, and in a wide variety of aircraft and locomotives).

Where very high power-to-weight ratios are required, internal combustion engines appear in the form of gas turbines. These applications include jet aircraft, helicopters, large ships and electric generators.

Types of internal combustion engine

At one time, the word, "Engine" meant any piece of machinery-a sense that persists in expressions such as siege engine. A "motor" (from Latin motor, "mover") is any machine that produces mechanical power. Traditionally, electric motors are not referred to as "Engines"; however, combustion engines are often referred to as "motors." (An electric engine refers to a locomotive operated by electricity.)

Engines can be classified in many different ways: By the engine cycle used, the layout of the engine, source of energy, the use of the engine, or by the cooling system employed.

Principles of operation

Reciprocating:

- Two-stroke cycle
- Four-stroke cycle
- Six-stroke engine
- Diesel engine
- Atkinson cycle

Rotary:

- Wankel engine

Continuous combustion:

Brayton cycle:

- Gas turbine
- Jet engine (including turbojet, turbofan, ramjet, Rocket etc..

ENGINE CONFIGURATIONS

Internal combustion engines can be classified by their configuration.

Four stroke configuration

Operation

1. Intake
2. Compression
3. Power
4. Exhaust

As their name implies, operation of four stroke internal combustion engines have four basic steps that repeat with every two revolutions of the engine:

1. Intake
   - Combustible mixtures are emplaced in the combustion chamber
2. Compression
   - The mixtures are placed under pressure
3. Combustion (Power)
   - The mixture is burnt, almost invariably a deflagration, although a few systems involve detonation. The hot mixture is expanded, pressing on and moving parts of the engine and performing useful work.
4. Exhaust
   - The cooled combustion products are exhausted into the atmosphere

Many engines overlap these steps in time; jet engines do all steps simultaneously at different parts of the engines.

Combustion

All internal combustion engines depend on the exothermic chemical process of combustion: the reaction of a fuel, typically with oxygen from the air (though it is possible to inject nitrous oxide in order to do more of the same thing and gain a power boost). The combustion process typically results in the production of a great quantity of heat, as well as the production of steam and carbon dioxide and other chemicals at very high temperature; the temperature reached is determined by the chemical make up of the fuel and oxidisers, as well as by the compression and other factors.

The most common modern fuels are made up of hydrocarbons and are derived mostly from fossil fuels (petroleum). Fossil fuels include diesel fuel, gasoline and petroleum gas, and the rarer use of propane. Except for the fuel delivery components, most internal combustion engines that are designed for gasoline use can run on natural gas or liquefied petroleum gases without major modifications. Large diesels can run with air mixed with gases and a pilot diesel fuel ignition injection.

Liquid and gaseous biofuels, such as ethanol and biodiesel (a form of diesel fuel that is produced from crops that yield triglycerides such as soybean oil), can also be used. Engines with appropriate modifications can also run on hydrogen gas, wood gas, or charcoal gas, as well as from so-called producer gas made from other convenient biomass.

Internal combustion engines require ignition of the mixture, either by spark ignition (SI) or compression ignition (CI). Before the invention of reliable electrical methods, hot tube and flame methods were used.

GASOLINE IGNITION PROCESS

Gasoline engine ignition systems generally rely on a combination of a lead-acid battery and an induction coil to provide a high-voltage electric spark to ignite the air-fuel mix in the engine's cylinders. This battery is recharged during operation using an electricity-generating device such as an alternator or generator driven by the engine. Gasoline engines take in a mixture of air and gasoline and compress it to not more than 12.8 bar (1.28 MPa), then use a spark plug to ignite the mixture when it is compressed by the piston head in each cylinder.

Diesel Ignition Process

Diesel engines and HCCI (Homogeneous charge compression ignition) engines, rely solely on heat and pressure created by the engine in its compression process for ignition. The compression level that occurs is usually twice or more than a gasoline engine. Diesel engines will take in air only, and shortly before peak compression, a small quantity of diesel fuel is sprayed into the cylinder via a fuel injector that allows the fuel to instantly ignite.

HCCI type engines will take in both air and fuel but continue to rely on an unaided auto-combustion process, due to higher pressures and heat. This is also why diesel and HCCI engines are more susceptible to cold-starting issues, although they will run just as well in cold weather once started. Light duty diesel engines with indirect injection in automobiles and light trucks employ glowplugs that pre-heat the combustion chamber just before starting to reduce no-start conditions in cold weather.

Most diesels also have a battery and charging system; nevertheless, this system is secondary and is added by manufacturers as a luxury for the ease of starting, turning fuel on and off (which can also be done via a switch or mechanical apparatus), and for running auxiliary electrical components and accessories. Most new engines rely on electrical and electronic engine control units (ECU) that also adjust the combustion process to increase efficiency and reduce emissions.

Two stroke configuration

Engines based on the two-stroke cycle use two strokes (one up, one down) for every power stroke. Since there are no dedicated intake or exhaust strokes, alternative methods must be used to scavenge the cylinders. The most common method in spark-ignition two-strokes is to use the downward motion of the piston to pressurize fresh charge in the crankcase, which is then blown through the cylinder through ports in the cylinder walls.

Spark-ignition two-strokes are small and light for their power output and mechanically very simple; however, they are also generally less efficient and more polluting than their four-stroke counterparts. In terms of power per cm^3, a two-stroke engine produces comparable power to an equivalent four-stroke engine. The advantage of having one power stroke for every 360° of crankshaft rotation (compared to 720° in a 4 stroke motor) is balanced by the less complete intake and exhaust and the shorter effective compression and power strokes. It may be possible for a two stroke to produce more power than an equivalent four stroke, over a narrow range of engine speeds, at the expense of less power at other speeds.

Small displacement, crankcase-scavenged two-stroke engines have been less fuel-efficient than other types of engines when the fuel is mixed with the air prior to scavenging allowing some of it to escape out of the exhaust port. Modern designs (Sarich and Paggio) use air-assisted fuel injection which avoids this loss, and are more efficient than comparably sized four-stroke engines. Fuel injection is essential for a modern two-stroke engine in order to meet ever more stringent emission standards.

Research continues into improving many aspects of two-stroke motors including direct fuel injection, amongst other things. The initial results have produced motors that are much cleaner burning than their traditional counterparts. Two-stroke engines are widely used in snowmobiles, lawnmowers, string trimmers, chain saws, jet skis, mopeds, outboard motors, and many motorcycles. Two-stroke engines have the advantage of an increased specific power ratio (i.e. power to volume ratio), typically around 1.5 times that of a typical four-stroke engine.

The largest internal combustion engines in the world are two-stroke diesels, used in some locomotives and large ships. They use forced induction (similar to super-charging, or turbocharging) to scavenge the cylinders; an example of this type of motor is the Wartsila-Sulzer turbocharged two-stroke diesel as used in large container ships. It is the most efficient and powerful internal combustion engine in the world with over 50% thermal efficiency. For comparison, the most efficient small four-stroke motors are around 43% thermal efficiency (SAE 900648); size is an advantage for efficiency due to the increase in the ratio of volume to surface area.

Common cylinder configurations include the straight or inline configuration, the more compact V configuration, and the wider but smoother flat or boxer configuration. Aircraft engines can also adopt a radial configuration which allows more effective cooling. More unusual configurations such as the H, U, X, and W have also been used.

Multiple crankshaft configurations do not necessarily need a cylinder head at all because they can instead have a piston at each end of the cylinder called an opposed piston design. Because here gas in- and outlets are positioned at opposed ends of the cylinder, one can achieve uniflow scavenging, which is, like in the four stroke engine, efficient over a wide range of revolution numbers. Also the thermal efficiency is improved because of lack of cylinder heads. This design was used in the Junkers Jumo 205 diesel aircraft engine, using at either end of a single bank of cylinders with two crankshafts, and most remarkably in the Napier Deltic diesel engines. These used three crankshafts to serve three banks of double-ended cylinders arranged in an equilateral triangle with the crankshafts at the corners. It was also used in single-bank locomotive engines, and continues to be used for marine engines, both for propulsion and for auxiliary generators.

Wankel

The Wankel engine (rotary engine) does not have piston strokes. It operates with the same separation of phases as the four-stroke engine with the phases taking place in separate locations in the engine. In thermodynamic terms it follows the Otto engine cycle, so may be thought of as a "four-phase" engine.

While it is true that three power strokes typically occur per rotor revolution due to the 3:1 revolution ratio of the rotor to the eccentric shaft, only one power stroke per shaft revolution actually occurs; this engine provides three power 'strokes' per revolution per rotor giving it a greater power-to-weight ratio than piston engines. This type of engine is most notably used in the current Mazda RX-8, the earlier RX-7, and other models.

Wave disk engine

The Wave disk engine is an internal combustion engine which does away with pistons, crankshafts and valves, and replaces them with a disc-shaped shock wave generator. Compression is achieved through the generation of shock waves in a spinning air fuel mixture.

The first prototype was demonstrated in March 2011 by a team at Michigan State University. It promises to be 3.5 times more efficient, 20% lighter, 30% cheaper to manufacture, and achieve a 90% reduction in emissions when compared to conventional internal combustion engine designs.

Gas turbines

A gas turbine is a rotary machine similar in principle to a steam turbine and it consists of three main components: a compressor, a combustion chamber, and a turbine. The air after being compressed in the compressor is heated by burning fuel in it. About ? of the heated air combined with the products of combustion is expanded in a turbine resulting in work output which is used to drive the compressor. The rest (about ?) is available as useful work output.

Jet engine

Jet engines take a large volume of hot gas from a combustion process (typically a gas turbine, but rocket forms of jet propulsion often use solid or liquid propellants, and ramjet forms also lack the gas turbine) and feed it through a nozzle which accelerates the jet to high speed. As the jet accelerates through the nozzle, this creates thrust and in turn does useful work.

Two-stroke

This system manages to pack one power stroke into every two strokes of the piston (up-down). This is achieved by exhausting and re-charging the cylinder simultaneously.

The steps involved here are:

1. Intake and exhaust occur at bottom dead center. Some form of pressure is needed, either crankcase compression or super-charging.
2. Compression stroke: Fuel-air mix compressed and ignited. In case of Diesel: Air compressed, fuel injected and self ignited
3. Power stroke: piston is pushed downwards by the hot exhaust gases.

Two Stroke Spark Ignition (SI) engine:

In a two strokes SI engine a cycle is completed in two stroke of a piston or one complete revolution (360°) of a crankshaft. In this engine the suction stroke and exhaust strokes are eliminated and ports are used instead of valves. Petrol is used in this type of stupidity The major components of a two stroke spark Ignition engine are: Cylinder: It is a cylindrical vessel in which a piston makes an up and down motion. Piston: It is a cylindrical component making an up and down movement in the cylinder.

Combustion Chamber: It is the portion above the cylinder in which the combustion of the fuel-air mixture takes place. Inlet and exhaust ports: The inlet port allows the fresh fuel-air mixture to enter the combustion chamber and the exhaust port discharges the products of combustion. Crank shaft: a shaft which converts the reciprocating motion of piston into the rotary motion. Connecting rod: connects the piston with the crankshaft. Cam shaft: The cam shaft controls the opening and closing of inlet and Exhaust valves. Spark plug: located at the cylinder head. It is used to initiate the combustion process.

Working: When the piston moves from bottom dead centre to top dead centre, the fresh air and fuel mixture enters the crank chamber through the valve. The mixture enters due to the pressure difference between the crank chamber and outer atmosphere. At the same time the fuel-air mixture above the piston is compressed.

Ignition with the help of spark plug takes place at the end of stroke. Due to the explosion of the gases, the piston moves downward. When the piston moves downwards the valve closes and the fuel-air mixture inside the crank chamber is compressed. When the piston is at the bottom dead centre, the burnt gases escape from the exhaust port.

At the same time the transfer port is uncovered and the compressed charge from the crank chamber enters into the combustion chamber through transfer port. This fresh charge is deflected upwards by a hump provided on the top of the piston. This fresh charge removes the exhaust gases from the combustion chamber. Again the piston moves from bottom dead centre to top dead centre and the fuel-air mixture gets compressed when the both the Exhaust port and Transfer ports are covered. The cycle is repeated.

Four-stroke

Engines based on the four-stroke ("Otto cycle") have one power stroke for every four strokes (up-down-up-down) and employ spark plug ignition. Combustion occurs rapidly, and during combustion the volume varies little ("constant volume").

They are used in cars, larger boats, some motorcycles, and many light aircraft. They are generally quieter, more efficient, and larger than their two-stroke counterparts.

The steps involved here are:

1. Intake stroke: Air and vaporized fuel are drawn in.
2. Compression stroke: Fuel vapor and air are compressed and ignited.
3. Combustion stroke: Fuel combusts and piston is pushed downwards.
4. Exhaust stroke: Exhaust is driven out. During the 1st, 2nd, and 4th stroke the piston is relying on power and the momentum generated by the other pistons. In that case, a four-cylinder engine would be less powerful than a six or eight cylinder engine.

There are a number of variations of these cycles, most notably the Atkinson and Miller cycles. The diesel cycle is somewhat different.

Diesel cycle

Most truck and automotive diesel engines use a cycle reminiscent of a four-stroke cycle, but with a compression heating ignition system, rather than needing a separate ignition system. This variation is called the diesel cycle. In the diesel cycle, diesel fuel is injected directly into the cylinder so that combustion occurs at constant pressure, as the piston moves.

Five-stroke

The British company ILMOR presented a prototype of 5-Stroke double expansion engine, having two outer cylinders, working as usual, plus a central one, larger in diameter, that performs the double expansion of exhaust gas from the other cylinders, with an increased efficiency in the gas energy use, and an improved SFC. This engine corresponds to a 2003 US patent by Gerhard Schmitz, and was developed apparently also by Honda of Japan for a Quad engine.

This engine has a similar precedent in an Spanish 1942 patent (# P0156621), by Francisco Jimeno-Cataneo, and a 1975 patent (# P0433850) by Carlos Ubierna-Laciana (www.oepm.es). The concept of double expansion was developed early in the history of ICE by Otto himself, in 1879, and a Connecticut (USA) based company, EHV, built in 1906 some engines and cars with this principle, that didn't give the expected results.

Six-stroke

First invented in 1883, the six-stroke engine has seen renewed interest over the last 20 or so years. Four kinds of six-stroke use a regular piston in a regular cylinder

(Griffin six-stroke, Bajulaz six-stroke, Velozeta six-stroke and Crower six-stroke), firing every three crankshaft revolutions. The systems capture the wasted heat of the four-stroke Otto cycle with an injection of air or water. The Beare Head and "piston charger" engines operate as opposed-piston engines, two pistons in a single cylinder, firing every two revolutions rather more like a regular four-stroke.

Brayton cycle

A gas turbine is a rotary machine somewhat similar in principle to a steam turbine and it consists of three main components: a compressor, a combustion chamber, and a turbine. The air after being compressed in the compressor is heated by burning fuel in it, this heats and expands the air, and this extra energy is tapped by the turbine which in turn powers the compressor closing the cycle and powering the shaft.

Gas turbine cycle engines employ a continuous combustion system where compression, combustion, and expansion occur simultaneously at different places in the engine-giving continuous power. Notably, the combustion takes place at constant pressure, rather than with the Otto cycle, constant volume.

Obsolete

The very first internal combustion engines did not compress the mixture. The first part of the piston downstroke drew in a fuel-air mixture, then the inlet valve closed and, in the remainder of the down-stroke, the fuel-air mixture fired. The exhaust valve opened for the piston upstroke. These attempts at imitating the principle of a steam engine were very inefficient.

Fuels and oxidizers

Engines are often classified by the fuel (or propellant) used.

Fuels

Nowadays, fuels used include:

- Petroleum:
  - Petroleum spirit (North American term: gasoline, British term: petrol)
  - Petroleum diesel.
  - Autogas (liquified petroleum gas).
  - Compressed natural gas.
  - Jet fuel (aviation fuel)
  - Residual fuel

- Coal:
  - o Most methanol is made from coal.
  - o Gasoline can be made from carbon (coal) using the Fischer-Tropsch process
  - o Diesel fuel can be made from carbon using the Fischer-Tropsch process
- Biofuels and vegoils:
  - o Peanut oil and other vegoils.
  - o Biofuels:
- Biobutanol (replaces gasoline).
- Biodiesel (replaces petrodiesel).
- Bioethanol and Biomethanol (wood alcohol) and other biofuels.
- Biogas
- Hydrogen (mainly spacecraft rocket engines)

Even fluidized metal powders and explosives have seen some use. Engines that use gases for fuel are called gas engines and those that use liquid hydrocarbons are called oil engines, however gasoline engines are also often colloquially referred to as, "gas engines" ("petrol engines" in the UK).

The main limitations on fuels are that it must be easily transportable through the fuel system to the combustion chamber, and that the fuel releases sufficient energy in the form of heat upon combustion to make practical use of the engine.

Diesel engines are generally heavier, noisier, and more powerful at lower speeds than gasoline engines. They are also more fuel-efficient in most circumstances and are used in heavy road vehicles, some automobiles (increasingly so for their increased fuel efficiency over gasoline engines), ships, railway locomotives, and light aircraft.

Gasoline engines are used in most other road vehicles including most cars, motorcycles, and mopeds. Note that in Europe, sophisticated diesel-engined cars have taken over about 40% of the market since the 1990s. There are also engines that run on hydrogen, methanol, ethanol, liquefied petroleum gas (LPG), biodiesel, wood gas, & charcoal gas. Paraffin and tractor vaporizing oil (TVO) engines are no longer seen.

Hydrogen

Hydrogen could eventually replace conventional fossil fuels in traditional internal combustion engines. Alternatively fuel cell technology may come to deliver its promise and the use of the internal combustion engines could even be phased out.

Although there are multiple ways of producing free hydrogen, those methods require converting combustible molecules into hydrogen or consuming electric energy. Unless that electricity is produced from a renewable source-and is not required for other purposes- hydrogen does not solve any energy crisis. In many situations, the disadvantage of hydrogen, relative to carbon fuels, is its storage. Liquid hydrogen has extremely low density (14 times lower than water) and requires extensive insulation-whilst gaseous hydrogen requires heavy tankage.

Even when liquefied, hydrogen has a higher specific energy but the volumetric energetic storage is still roughly five times lower than petrol. However the energy density of hydrogen is considerably higher than that of electric batteries, making it a serious contender as an energy carrier to replace fossil fuels. The 'Hydrogen on Demand' process creates hydrogen as it is needed, but has other issues such as the high price of the sodium borohydride which is the raw material.

Oxidizers

Since air is plentiful at the surface of the earth, the oxidizer is typically atmospheric oxygen which has the advantage of not being stored within the vehicle, increasing the power-to-weight and power to volume ratios. There are other materials that are used for special purposes, often to increase power output or to allow operation under water or in space.

- Compressed air has been commonly used in torpedoes.
- Compressed oxygen, as well as some compressed air, was used in the Japanese Type 93 torpedo. Some submarines are designed to carry pure oxygen. Rockets very often use liquid oxygen.
- Nitromethane is added to some racing and model fuels to increase power and control combustion.
- Nitrous oxide has been used-with extra gasoline-in tactical aircraft and in specially equipped cars to allow short bursts of added power from engines that otherwise run on gasoline and air. It is also used in the Burt Rutan rocket spacecraft.
- Hydrogen peroxide power was under development for German World War II submarines and may have been used in some non-nuclear submarines

and was used on some rocket engines (notably Black Arrow and Me-163 rocket plane)

- Other chemicals such as chlorine or fluorine have been used experimentally, but have not been found to be practical.

Engine starting

An internal combustion engine is not usually self-starting so an auxiliary machine is required to start it. Many different systems have been used in the past but modern engines are usually started by an electric motor in the small and medium sizes or by compressed air in the large sizes.

Measures of engine performance

Engine types vary greatly in a number of different ways:

- Energy efficiency
- Fuel/propellant consumption (brake specific fuel consumption for shaft engines, thrust specific fuel consumption for jet engines)
- Power to weight ratio
- Thrust to weight ratio
- Torque curves (for shaft engines) thrust lapse (jet engines)
- Compression ratio for piston engines, Overall pressure ratio for jet engines and gas turbines

Energy efficiency

Once ignited and burnt, the combustion products-hot gases-have more available thermal energy than the original compressed fuel-air mixture (which had higher chemical energy). The available energy is manifested as high temperature and pressure that can be translated into work by the engine. In a reciprocating engine, the high-pressure gases inside the cylinders drive the engine's pistons.

Once the available energy has been removed, the remaining hot gases are vented (often by opening a valve or exposing the exhaust outlet) and this allows the piston to return to its previous position (top dead center, or TDC). The piston can then proceed to the next phase of its cycle, which varies between engines. Any heat that isn't translated into work is normally considered a waste product and is removed from the engine either by an air or liquid cooling system.

Engine efficiency can be discussed in a number of ways but it usually involves a comparison of the total chemical energy in the fuels, and the useful energy extracted from the fuels in the form of kinetic energy. The most fundamental and abstract discussion of engine efficiency is the thermodynamic limit for extracting energy from the fuel defined by a thermodynamic cycle. The most comprehensive is the empirical fuel efficiency of the total engine system for accomplishing a desired task; for example, the miles per gallon accumulated.

Internal combustion engines are primarily heat engines and as such the phenomenon that limits their efficiency is described by thermodynamic cycles. None of these cycles exceed the limit defined by the Carnot cycle which states that the overall efficiency is dictated by the difference between the lower and upper operating temperatures of the engine. A terrestrial engine is usually and fundamentally limited by the upper thermal stability derived from the material used to make up the engine.

All metals and alloys eventually melt or decompose and there is significant researching into ceramic materials that can be made with higher thermal stabilities and desirable structural properties. Higher thermal stability allows for greater temperature difference between the lower and upper operating temperatures-thus greater thermodynamic efficiency.

The thermodynamic limits assume that the engine is operating in ideal conditions: a frictionless world, ideal gases, perfect insulators, and operation at infinite time. The real world is substantially more complex and all the complexities reduce the efficiency. In addition, real engines run best at specific loads and rates as described by their power band. For example, a car cruising on a highway is usually operating significantly below its ideal load, because the engine is designed for the higher loads desired for rapid acceleration.

The applications of engines are used as contributed drag on the total system reducing overall efficiency, such as wind resistance designs for vehicles. These and many other losses result in an engine's real-world fuel economy that is usually measured in the units of miles per gallon (or fuel consumption in liters per 100 kilometers) for automobiles. The miles in miles per gallon represents a meaningful amount of work and the volume of hydrocarbon implies a standard energy content.

Most steel engines have a thermodynamic limit of 37%. Even when aided with turbochargers and stock efficiency aids, most engines retain an average efficiency of about 18%-20%. Rocket engine efficiencies are better still, up to 70%, because they combust at very high temperatures and pressures and are able to have very high expansion ratios.

There are many inventions concerned with increasing the efficiency of IC engines. In general, practical engines are always compromised by trade-offs between different properties such as efficiency, weight, power, heat, response, exhaust emissions, or noise. Sometimes economy also plays a role in not only the cost of manufacturing the engine itself, but also manufacturing and distributing the fuel. Increasing the engine's efficiency brings better fuel economy but only if the fuel cost per energy content is the same.

Measures of fuel/propellant efficiency

For stationary and shaft engines including propeller engines, fuel consumption is measured by calculating the brake specific fuel consumption which measures the mass flow rate of fuel consumption divided by the power produced.

For internal combustion engines in the form of jet engines, the power output varies drastically with airspeed and a less variable measure is used: thrust specific fuel consumption (TSFC), which is the number of pounds of propellant that is needed to generate impulses that measure a pound force-hour. In metric units, the number of grams of propellant needed to generate an impulse that measures one kilonewton-second.

For rockets, TSFC can be used, but typically other equivalent measures are traditionally used, such as specific impulse and effective exhaust velocity.

AIR AND NOISE POLLUTION

Air pollution

Internal combustion engines such as reciprocating internal combustion engines produce air pollution emissions, due to incomplete combustion of carbonaceous fuel. The main derivatives of the process are carbon dioxide CO2, water and some soot - also called particulate matter (PM). The effects of inhaling particulate matter have been studied in humans and animals and include asthma, lung cancer, cardiovascular issues, and premature death. There are however some additional products of the combustion process that include nitrogen oxides and sulfur and some uncombusted hydrocarbons, depending on the operating conditions and the fuel-air ratio.

Not all of the fuel will be completely consumed by the combustion process; a small amount of fuel will be present after combustion, some of which can react to form oxygenates, such as formaldehyde or acetaldehyde, or hydrocarbons not initially present in the fuel mixture. The primary causes of this is the need to operate near the stoichiometric ratio for gasoline engines in order to achieve combustion and the resulting "quench" of the flame by the relatively cool cylinder walls, otherwise the fuel would burn more completely in excess air. When running at lower speeds, quenching is commonly observed in diesel (compression ignition) engines that run on natural gas.

It reduces the efficiency and increases knocking, sometimes causing the engine to stall. Increasing the amount of air in the engine reduces the amount of the first two pollutants, but tends to encourage the oxygen and nitrogen in the air to combine to produce nitrogen oxides (NOx) that has been demonstrated to be hazardous to both plant and animal health. Further chemicals released are benzene and 1,3-butadiene that are also particularly harmful; and not all of the fuel burns up completely, so carbon monoxide (CO) is also produced.

Carbon fuels contain sulfur and impurities that eventually lead to producing sulfur monoxides (SO) and sulfur dioxide (SO2) in the exhaust which promotes acid rain. One final element in exhaust pollution is ozone (O3). This is not emitted directly but made in the air by the action of sunlight on other pollutants to form "ground level ozone", which, unlike the "ozone layer" in the high atmosphere, is regarded as a bad thing if the levels are too high. Ozone is broken down by nitrogen oxides, so one tends to be lower where the other is higher.

For the pollutants described above (nitrogen oxides, carbon monoxide, sulphur dioxide, and ozone), there are accepted levels that are set by legislation to which no harmful effects are observed - even in sensitive population groups. For the other three: benzene, 1,3-butadiene, and particulates, there is no way of proving they are safe at any level so the experts set standards where the risk to health is, "exceedingly small".

Noise pollution

Significant contributions to noise pollution are made by internal combustion engines. Automobile and truck traffic operating on highways and street systems produce noise, as do aircraft flights due to jet noise, particularly supersonic-capable aircraft. Rocket engines create the most intense noise.

Idling

Internal combustion engines continue to consume fuel and emit pollutants when idling so it is desirable to keep periods of idling to a minimum. Many bus companies now instruct drivers to switch off the engine when the bus is waiting at a terminus.

In the UK (but applying only to England), the Road Traffic (Vehicle Emissions) (Fixed Penalty) Regulations 2002 (Statutory Instrument 2002 No. 1808) introduced the concept of a "stationary idling offence". This means that a driver can be ordered "by an authorised person... upon production of evidence of his authorisation, require him to stop the running of the engine of that vehicle" and a "person who fails to comply... shall be guilty of an offence and be liable on summary conviction to a fine not exceeding level 3 on the standard scale". Only a few local authorities have implemented the regulations, one of them being Oxford City Council.

EXTERNAL COMBUSTION ENGINE

An external combustion engine (EC engine) is a heat engine where an (internal) working fluid is heated by combustion of an external source, through the engine wall or a heat exchanger. The fluid then, by expanding and acting on the mechanism of the engine produces motion and usable work. The fluid is then cooled, compressed and reused (closed cycle), or (less commonly) dumped, and cool fluid pulled in (open cycle air engine).

"Combustion" refers to burning fuel with an oxidizer, to supply the heat. Engines of similar (or even identical) configuration and operation may use a supply of heat from other sources such as nuclear, solar, geothermal or exothermic reactions not involving combustion; but are not then strictly classed as external combustion engines, but as external thermal engines.

The working fluid can be a gas as in a Stirling engine, or steam as in a steam engine. The fluid can be of any composition; gas is by far the most common, although even single-phase liquid is sometimes used. In the case of the steam engine, or the Organic Rankine Cycle the fluid changes phases between liquid and gas.

Reciprocating Engine

A reciprocating engine, also often known as a piston engine, is a heat engine that uses one or more reciprocating pistons to convert pressure into a rotating motion. This article describes the common features of all types. The main types are: the internal combustion engine, used extensively in motor vehicles; the steam engine, the mainstay of the Industrial Revolution; and the niche application Stirling engine.

There may be one or more pistons. Each piston is inside a cylinder, into which a gas is introduced, either already hot and under pressure (steam engine), or heated inside the cylinder either by ignition of a fuel air mixture (internal combustion engine) or by contact with a hot heat exchanger in the cylinder (Stirling engine).

The hot gases expand, pushing the piston to the bottom of the cylinder. The piston is returned to the cylinder top (Top Dead Centre) either by a flywheel or the power from other pistons connected to the same shaft. In most types the expanded or "exhausted" gases are removed from the cylinder by this stroke. The exception is the Stirling engine, which repeatedly heats and cools the same sealed quantity of gas.

Components of a typical, four stroke cycle, internal combustion piston engine.

E - Exhaust camshaft

I - Intake camshaft

S - Spark plug

V - Valves

P - Piston

R - Connecting rod

C - Crankshaft

W - Water jacket for coolant flow

In some designs the piston may be powered in both directions in the cylinder in which case it is said to be double acting.

Steam piston engine

A labeled schematic diagram of a typical single cylinder, simple expansion, double-acting high pressure steam engine. Power takeoff from the engine is by way of a belt.

1. Piston
2. Piston rod
3. Crosshead bearing
4. Connecting rod
5. Crank
6. Eccentric valve motion
7. Flywheel
8. Sliding valve
9. Centrifugal governor.

In all types, the linear movement of the piston is converted to a rotating movement via a connecting rod and a crankshaft or by a swashplate. A flywheel is often used to ensure smooth rotation. The more cylinders a reciprocating engine has, generally, the more vibration-free (smoothly) it can operate. The power of a reciprocating engine is proportional to the volume of the combined pistons' displacement.

A seal needs to be made between the sliding piston and the walls of the cylinder so that the high pressure gas above the piston does not leak past it and reduce the efficiency of the engine. This seal is provided by one or more piston rings. These are rings made of a hard metal which are sprung into a circular groove in the piston head. The rings fit tightly in the groove and press against the cyinder wall to form a seal.

It is common for such engines to be classified by the number and alignment of cylinders and the total volume of displacement of gas by the pistons moving in the cylinders usually measured in cubic centimetres (cm^3 or cc) or litres (l) or (L) (US:liter). For example for internal combustion engines, single and two-cylinder designs are common in smaller vehicles such as motorcycles, while automobiles typically have between four and eight, and locomotives, and ships may have a dozen cylinders or more. Cylinder capacities may range from 10 cm^3 or less in model engines up to several thousand cubic centimetres in ships' engines.

The compression ratio is a measure of the performance in an internal-combustion engine or a Stirling Engine. It is the ratio between the volume of the cylinder, when the piston is at the bottom of its stroke, and the volume when the piston is at the top of its stroke.

The bore/stroke ratio is the ratio of the diameter of the piston, or "bore", to the length of travel within the cylinder, or "stroke". If this is around 1 the engine is said to be "square", if it is greater than 1, i.e. the bore is larger than the stroke, it is "oversquare". If it is less than 1, i.e. the stroke is larger than the bore, it is "undersquare".

Cylinders may be aligned in line, in a V configuration, horizontally opposite each other, or radially around the crankshaft. Opposed-piston engines put two pistons working at opposite ends of the same cylinder and this has been extended into triangular arrangements such as the Napier Deltic. Some designs have set the cylinders in motion around the shaft, see the Rotary engine.

Stirling piston engine

Rhombic Drive Beta Stirling Engine Design showing the second displacer piston (green) within the cylinder which shunts the working gas between the hot and cold ends, but produces no power itself.

Pink - Hot cylinder wall,

Dark grey - Cold cylinder wall,

Green - Displacer piston,

Dark blue - Power piston,

Light blue - Flywheels

In steam engines and internal combustion engines, valves are required to allow the entry and exit of gasses at the correct time in the piston's cycle. These are worked by cams or cranks driven by the shaft of the engine. Early designs used the D slide valve but this has been largely superseded by Piston valve or Poppet valve designs. In steam engines the point in the piston cycle at which the steam inlet valve closes is called the cutoff and this can often be controlled to adjust the torque supplied by the engine.

Internal combustion engines operate through a sequence of strokes which admit and remove gases to and from the cylinder. These operations are repeated cyclically and an engine is said to be 2-stroke, 4-stroke or 6-stroke depending on the number of strokes it takes to complete a cycle.

In some steam engines, the cylinders may be of varying size with the smallest bore cylinder working the highest pressure steam. This is then fed through one or more, increasingly larger bore cylinders successively, to extract power from the steam at increasingly lower pressures. These engines are called Compound engines.

An early known example of rotary to reciprocating motion can be found in a number of Roman saw mills (dating to the 3rd to 6th century AD) in which a crank and connecting rod mechanism converted the rotary motion of the waterwheel into the linear movement of the saw blades. Another early example was the reciprocating piston pump of Al-Jazari in 1206.

The reciprocating engine developed in Europe during the 18th century, first as the atmospheric engine then later as the steam engine. These were followed by the Stirling engine and internal combustion engine in the 19th century. Today the most common form of reciprocating engine is the internal combustion engine running on the combustion of petrol, diesel, Liquefied petroleum gas (LPG) or compressed natural gas (CNG) and used to power motor vehicles.

One of the most advanced reciprocating engines ever made was the 28-cylinder, 3,500 hp (2,600 kW) Pratt & Whitney R-4360 "Wasp Major" radial engine which powered the last generation of large piston-engined planes before the jet engine and turboprop took over from 1944 onward. It had a total engine capacity of 71.5 litres (2.52 cu ft).

The largest reciprocating engine in production at present, but not the largest ever built, is the Wärtsilä-Sulzer RTA96-C turbocharged two-stroke diesel engine of 2006 built by Japan's Diesel United, Ltd. It is used to power the largest modern container ships such as the Emma Mærsk. It is five stories high (13.5 m/44 ft), 27 metres (89 ft) long, and weighs over 2,300 metric tons (2,500 short tons) in its largest 14 cylinders version producing more than 84.42 MW (114,800 bhp). Each cylinder has a capacity of 1,820 litres (64 cu ft), making a total capacity of 25,480 litres (900 cu ft) for the largest versions.

Engine capacity

For piston engines, an engine's capacity is the engine displacement, in other words the volume swept by all the pistons of an engine in a single movement. It is generally measured in litres (L) or cubic inches (c.i.d. or cu in or in^3) for larger engines, and cubic centimetres (abbreviated cc) for smaller engines. Engines with greater capacities are more powerful and provide greater torque at lower speed (rpm) and consumption of fuel increases accordingly.

Other modern non-internal combustion types

Reciprocating engines that are powered by compressed air, steam or other hot gases are still used in some applications such as to drive many modern torpedoes or as pollution-free motive power. Most steam-driven applications use steam turbines, which are more efficient than piston engines.

The French-designed FlowAIR vehicles use compressed air stored in a cylinder to drive a reciprocating engine in a pollution-free urban vehicle.

Torpedoes may use a working gas produced by high test peroxide or Otto fuel II, which pressurise without combustion. The 230 kg (510 lb) Mark 46 torpedo, for example, can travel 11 km (6.8 mi) underwater at 74 km/h (46 mph) fuelled by Otto fuel without oxidant.

8

HYDRAULICS PRINCIPLES & WATER TURBINES

Hydraulics is a topic in applied science and engineering dealing with the mechanical properties of liquids. Fluid mechanics provides the theoretical foundation for hydraulics, which focuses on the engineering uses of fluid properties. In fluid power, hydraulics is used for the generation, control, and transmission of power by the use of pressurized liquids.

A simple hydraulic system consists of hydraulic fluid, pistons or rams, cylinders, accumulator or oil reservoir, a complete working mechanism, and safety devices. These systems are capable of remotely controlling a wide variety of equipment by transmitting force, carried by the hydraulic fluid, in a confined medium. Modern developments in hydraulics have involved many fields in engineering and transportation.

These systems transfer high forces rapidly and accurately even in small pipes of light weight, small size, any shape, and over a long distance. These systems play a vital role from small car's steering to super sonic aircraft's maneuvering devices. More powerful and accurate systems are also used in maneuvering huge ships.

Hydraulic topics range through most science and engineering disciplines, and cover concepts such as pipe flow, dam design, fluidics and fluid control circuitry, pumps, turbines, hydropower, computational fluid dynamics, flow measurement, river channel behavior and erosion.Free surface hydraulics is the branch of hydraulics dealing with frese surface flow, such as occurring in rivers, canals, lakes, estuaries and seas. Its sub-field open channel flow studies the flow in open channels.

Earlier, weights were lifted using pulleys, levers, block and tackles, etc. Movements for a ship's rudder or steering a vehicle where achieved by mechanical linkages like cams, levers, couplings, and gears which made the system complicated. These manual or mechanical methods of operation had several limitations. They also involved huge man power and long working hours for a particular job. As the population and technology increased exponentially, the demand for quicker and easier to operate equipment increased. To cater to this need, hydraulic machines were introduced.

A simple hydraulic system consists of hydraulic fluid, pistons or rams, cylinders, accumulator or oil reservoir, a complete working mechanism, and safety devices. These systems are capable of remotely controlling a wide variety of equipment by transmitting force, carried by the hydraulic fluid, in a confined medium. Modern developments in hydraulics have involved many fields in engineering and transportation. These systems transfer high forces rapidly and accurately even in small pipes of light weight, small size, any shape, and over a long distance. These systems play a vital role from small car's steering to super sonic aircraft's maneuvering devices. More powerful and accurate systems are also used in maneuvering huge ships.

ANCIENT AND MEDIEVAL ERA

Early uses of water power date back to Mesopotamia and ancient Egypt, where irrigation has been used since the 6th millennium BC and water clocks had been used since the early 2nd millennium BC. Other early examples of water power include the Qanat system in ancient Persia and the Turpan water system in ancient China.

Greek / Hellenistic world

Greeks continued and sophisticated the construction of water and hydraulic power systems. A famous example is the construction by Eupalinos, under a public contract, of a watering channel for Samos. An early example of the usage of hydraulic wheel, probably the earliest in Europe, is the Perachora wheel (3rd c. BC).

Notable is the construction of the first hydraulic automata by Ctesibius (flourished c. 270 BC) and Hero of Alexandria (c. 10-80 AD). Hero describes a number of working machines using hydraulic power, such as the force pump, which is known from many Roman sites as having been used for raising water and in fire engines.

China

In ancient China there was Sunshu Ao (6th century BC), Ximen Bao (5th century BC), Du Shi (circa 31 AD), Zhang Heng (78 - 139 AD), and Ma Jun (200 - 265 AD), while medieval China had Su Song (1020 - 1101 AD) and Shen Kuo (1031-1095). Du Shi employed a waterwheel to power the bellows of a blast furnace producing cast iron. Zhang Heng was the first to employ hydraulics to provide motive power in rotating an armillary sphere for astronomical observation.

Sri Lanka

In ancient Sri Lanka, hydraulics were widely used in the ancient kingdoms of Anuradhapura and Polonnaruwa. The discovery of the principle of the valve tower, or valve pit, for regulating the escape of water is credited to ingenuity more than 2,000 years ago. By the first century A.D, several large-scale irrigation works had been completed. Macro- and micro-hydraulics to provide for domestic horticultural and agricultural needs, surface drainage and erosion control, ornamental and recreational water courses and retaining structures and also cooling systems were in place in Sigiriya, Sri Lanka. The coral on the massive rock at the site includes cisterns for collecting water.

In Ancient Rome many different hydraulic applications were developed, including public water supplies, innumerable aqueducts, power using watermills and hydraulic mining. They were among the first to make use of the siphon to carry water across valleys, and used hushing on a large scale to prospect for and then extract metal ores. They used lead widely in plumbing systems for domestic and public supply, such as feeding thermae.

Hydraulic mining was used in the gold-fields of northern Spain, which was conquered by Augustus in 25 BC. The alluvial gold-mine of Las Medulas was one of the largest of their mines. It was worked by at least 7 long aqueducts, and the water streams were used to erode the soft deposits, and then wash the tailings for the valuable gold content.

Modern era (C. 1600-1870)

Benedetto Castelli

In 1619 Benedetto Castelli (1576 - 1578-1643), a student of Galileo Galilei, published the book Della Misura dell'Acque Correnti or "On the Measurement of Running Waters", one of the foundations of modern hydrodynamics. He served as a chief consultant to the Pope on hydraulic projects, i.e., management of rivers in the Papal States, beginning in 1626.

Blaise Pascal

Blaise Pascal (1623-1662-1672) studied fluid hydrodynamics and hydrostatics, centered on the principles of hydraulic fluids. His inventions include the hydraulic press, which multiplied a smaller force acting on a larger area into the application of a larger force totaled over a smaller area, transmitted through the same pressure (or same change of pressure) at both locations.

Pascal's law or principle states that for an incompressible fluid at rest, the difference in pressure is proportional to the difference in height and this difference remains the same whether or not the overall pressure of the fluid is changed by applying an external force. This implies that by increasing the pressure at any point in a confined fluid, there is an equal increase at every other point in the container, i.e., any change in pressure applied at any point of the fluid is transmitted undiminished throughout the fluids.

Jean Louis Marie Poiseuille

A French physician, Poiseuille researched the flow of blood through the body and discovered an important law governing the rate of flow with the diameter of the tube in which flow occurred.

Pascal's law

In the physical sciences, Pascal's law or the Principle of transmission of fluid-pressure states that "pressure exerted anywhere in a confined incompressible fluid is transmitted equally in all directions throughout the fluid such that the pressure ratio (initial difference) remains the same."

where

DP is the hydrostatic pressure (given in pascals in the SI system), or the difference in pressure at two points within a fluid column, due to the weight of the fluid;

r is the fluid density (in kilograms per cubic meter in the SI system);

g is acceleration due to gravity (normally using the sea level acceleration due to Earth's gravity in metres per second squared);

Dh is the height of fluid above the point of measurement, or the difference in elevation between the two points within the fluid column (in metres in SI).

The intuitive explanation of this formula is that the change in pressure between two elevations is due to the weight of the fluid between the elevations.

Note that the variation with height does not depend on any additional pressures. Therefore Pascal's law can be interpreted as saying that any change in pressure applied at any given point of the fluid is transmitted undiminished throughout the fluid. Equation: (P1)(V1) = (P2)(V2)

Applications

- The siphon
- The underlying principle of the hydraulic press
- Used for amplifying the force of the driver's foot in the braking system of most cars and trucks.
- Used in artesian wells, water towers, and dams.
- cuba divers must understand this principle. At a depth of 10 meters under water, pressure is twice the atmospheric pressure at sea level, and increases by about 105 kPa for each increase of 10 m depth.

Fluid statics

Fluid statics (also called hydrostatics) is the science of fluids at rest, and is a sub-field within fluid mechanics. The term usually refers to the mathematical treatment of the subject. It embraces the study of the conditions under which fluids are at rest in stable equilibrium. The use of fluid to do work is called hydraulics, and the science of fluids in motion is fluid dynamics.

Due to the fundamental nature of fluids, a fluid cannot remain at rest under the presence of a shear stress. However, fluids can exert pressure normal to any contacting surface. If a point in the fluid is thought of as an infinitesimally small cube, then it follows from the principles of equilibrium that the pressure on every side of this unit of fluid must be equal. If this were not the case, the fluid would move in the direction of the resulting force.

Thus, the pressure on a fluid at rest is isotropic; i.e., it acts with equal magnitude in all directions. This characteristic allows fluids to transmit force through the length of pipes or tubes; i.e., a force applied to a fluid in a pipe is transmitted, via the fluid, to the other end of the pipe.

This concept was first formulated, in a slightly extended form, by the French mathematician and philosopher Blaise Pascal in 1647 and would later be known as Pascal's law. This law has many important applications in hydraulics.

Hydrostatic pressure

Hydrostatic pressure is the pressure exerted by a fluid at equilibrium due to the force of gravity. A fluid in this condition is known as a hydrostatic fluid.

The hydrostatic pressure can be determined from a control volume analysis of an infinitesimally small cube of fluid. Since pressure is defined as the force exerted on a test area (p = F/A, with p: pressure, F: force normal to area A, A: area), and the only force acting on any such small cube of fluid is the weight of the fluid column above it, hydrostatic pressure can be calculated according to the following formula:

$$p(z) = \frac{1}{A}\int_{z0}^{z} dz' \iint_A dx'dy' \rho(z')g(z') = \int_{z0}^{z} dz' \rho(z')g(z')$$

where:

- r is the hydrostatic pressure (Pa),
- r is the fluid density (kg/m3),
- g is gravitational acceleration (m/s2),
- A is the test area (m2),
- z is the height (parallel to the direction of gravity) of the test area (m),
- z0 is the height of the zero reference point of the pressure (m).

For water and other liquids, this integral can be simplified significantly for many practical applications, based on the following two assumptions: Since many liquids can be considered incompressible, a reasonably good estimation can be made from assuming a constant density throughout the liquid.

(The same assumption cannot be made within a gaseous environment.) Also, since the height h of the fluid column between z and z0 is often reasonably small compared to the radius of the Earth, one can neglect the variation of g. Under these circumstances, the integral boils down to the simple formula:$p = pgh$

where h is the height z-z0 of the liquid column between the test volume and the zero reference point of the pressure. Note that this reference point should lie at or below the surface of the liquid. Otherwise, one has to split the integral into two (or more) terms with the constant ?liquid and ?(z')above. For example, the absolute pressure compared to vacuum is : $p = \rho g\, H + patm$

where H is the total height of the liquid column above the test area the surface, and patm is the atmospheric pressure, i.e., the pressure calculated from the remaining integral over the air column from the liquid surface to infinity.

Hydrostatic pressure has been used in the preservation of foods in a process called pascalization.

Atmospheric pressure

Statistical mechanics shows that, for a gas of constant temperature, T, its pressure, p will vary with height, h, as: $p(h) = p(0)e^{-Mgh/kT}$

where:

g = the acceleration due to gravity

T = Absolute temperature

k = Boltzmann constant

M = mass of a single molecule of gas

p = pressure

h = height

If there are multiple types of molecules in the gas, the partial pressure of each type will be given by this equation. Under most conditions, the distribution of each species of gas is independent of the other species.

Buoyancy

Any body of arbitrary shape which is immersed, partly or fully, in a fluid will experience the action of a net force in the opposite direction of the local pressure gradient. If this pressure gradient arises from gravity, the net force is in the vertical direction opposite that of the gravitational force. This vertical force is termed buoyancy or buoyant force and is equal in magnitude, but opposite in direction, to the weight of the displaced fluid.

In the case of a ship, for instance, its weight is balanced by shear force from the displaced water, allowing it to float. If more cargo is loaded onto the ship, it would sink more into the water - displacing more water and thus receive a higher buoyant force to balance the increased weight.

LIQUIDS-FLUIDS WITH FREE SURFACES

Liquids can have free surfaces at which they interface with gases, or with a vacuum. In general, the lack of the ability to sustain a shear stress entails that free surfaces rapidly adjust towards an equilibrium. However, on small length scales, there is an important balancing force from surface tension.

Capillary action

When liquids are constrained in vessels whose dimensions are small, compared to the relevant length scales, surface tension effects become important leading to the formation of a meniscus through capillary action. This capillary action has profound consequences for biological systems as it is part of one of the two driving mechanisms of the flow of water in plant xylem, the transpirational pull.

Drops

Without surface tension, drops would not be able to form. The dimensions and stability of drops are determined by surface tension. The drop's surface tension is directly proportional to the cohesion property of the fluid.

Partial Pressure

In a mixture of ideal gases, each gas has a partial pressure which is the pressure which the gas would have if it alone occupied the volume. The total pressure of a gas mixture is the sum of the partial pressures of each individual gas in the mixture.

In chemistry, the partial pressure of a gas in a mixture of gases is defined as above. The partial pressure of a gas dissolved in a liquid is the partial pressure of that gas which would be generated in a gas phase in equilibrium with the liquid at the same temperature. The partial pressure of a gas is a measure of thermodynamic activity of the gas's molecules.

Gases will always flow from a region of higher partial pressure to one of lower pressure; the larger this difference, the faster the flow. Gases dissolve, diffuse, and react according to their partial pressures, and not necessarily according to their concentrations in a gas mixture.

DALTON'S LAW OF PARTIAL PRESSURES

The partial pressure of an ideal gas in a mixture is equal to the pressure it would exert if it occupied the same volume alone at the same temperature. This is because ideal gas molecules are so far apart that they don't interfere with each other at all. Actual real-world gases come very close to this ideal.

A consequence of this is that the total pressure of a mixture of ideal gases is equal to the sum of the partial pressures of the individual gases in the mixture as stated by Dalton's law. For example, given an ideal gas mixture of nitrogen (N2), hydrogen (H2) and ammonia (NH3): $P = P_{N2} + P_{H2} + P_{NH3}$

where:

P = total pressure of the gas mixture

PN2 = partial pressure of nitrogen (N2)

PH2= partial pressure of hydrogen (H2)

PNH3= partial pressure of ammonia (NH3)

Ideal gas mixtures

Ideally the ratio of partial pressures is the same as the ratio of molecules. That is, the mole fraction of an individual gas component in an ideal gas mixture can be expressed in terms of the component's partial pressure or the moles of the component: $x_i = \frac{P_i}{P} = \frac{n_i}{n}$

and the partial pressure of an individual gas component in an ideal gas can be obtained using this expression: $P_i = x_i P$

where:

xi = mole fraction of any individual gas component in a gas mixture

Pi = partial pressure of any individual gas component in a gas mixture

ni = moles of any individual gas component in a gas mixture

n = total moles of the gas mixture

P = total pressure of the gas mixture

The mole fraction of a gas component in a gas mixture is equal to the volumetric fraction of that component in a gas mixture.

Partial volume(Amagat's law of additive volume)

The partial volume of a particular gas is the volume which the gas would have if it alone occupied the volume, with unchanged pressure and temperature, and is useful in gas mixtures, e.g. air, to focus on one particular gas component, e.g. oxygen.

It can be approximated both from partial pressure and molar fraction:

$$V_x = V_{tot} \times \frac{P_x}{P_{tot}} = V_{tot} \times \frac{n_x}{n_{tot}}$$

- Vx is the partial volume of any individual gas component (X)
- Vtot is the total volume in gas mixture
- Px is the partial pressure of gas X
- Ptot is the total pressure in gas mixture
- nx is the amount of substance of a gas (X)
- ntot is the total amount of substance in gas mixture.

Vapor pressure

Vapor pressure is the pressure of a vapor in equilibrium with its non-vapor phases (i.e., liquid or solid). Most often the term is used to describe a liquid's tendency to evaporate. It is a measure of the tendency of molecules and atoms to escape from a liquid or a solid. A liquid's atmospheric pressure boiling point corresponds to the temperature at which its vapor pressure is equal to the surrounding atmospheric pressure and it is often called the normal boiling point.

The higher the vapor pressure of a liquid at a given temperature, the lower the normal boiling point of the liquid.

The vapor pressure chart to the right has graphs of the vapor pressures versus temperatures for a variety of liquids. As can be seen in the chart, the liquids with the highest vapor pressures have the lowest normal boiling points.

For example, at any given temperature, propane has the highest vapor pressure of any of the liquids in the chart. It also has the lowest normal boiling point (-43.7 °C), which is where the vapor pressure curve of propane (the purple line) intersects the horizontal pressure line of one atmosphere (atm) of absolute vapor pressure.

Equilibrium constants of reactions involving gas mixtures

It is possible to work out the equilibrium constant for a chemical reaction involving a mixture of gases given the partial pressure of each gas and the overall reaction formula. For a reversible reaction involving gas reactants and gas products, such as: aA + bB ↔ cC + dD

the equilibrium constant of the reaction would be: $K_P = \frac{P_C^c P_D^d}{P_A^a P_B^b}$

where:

KP = the equilibrium constant of the reaction

a = coefficient of reactant A

b = coefficient of reactant B

c = coefficient of product C

d = coefficient of product D

P^c_C = the partial pressure of C raised to the power of c

P^d_D = the partial pressure of D raised to the power of d P^a_A

= the partial pressure of A raised to the power of a P^b_B

= the partial pressure of B raised to the power of b

For reversible reactions, changes in the total pressure, temperature or reactant concentrations will shift the equilibrium so as to favor either the right or left side of the reaction in accordance with Le Chatelier's Principle. However, the reaction kinetics may either oppose or enhance the equilibrium shift. In some cases, the reaction kinetics may be the over-riding factor to consider.

HENRY'S LAW AND THE SOLUBILITY OF GASES

Gases will dissolve in liquids to an extent that is determined by the equilibrium between the undissolved gas and the gas that has dissolved in the liquid (called the solvent). The equilibrium constant for that equilibrium is: $k = \frac{P_X}{C_X}$

where:

k = the equilibrium constant for the solvation process

PX = partial pressure of gas X in equilibrium with a solution containing some of the gas

CX = the concentration of gas X in the liquid solution

The form of the equilibrium constant shows that the concentration of a solute gas in a solution is directly proportional to the partial pressure of that gas above the solution. This statement is known as Henry's Law and the equilibrium constant k is quite often referred to as the Henry's Law constant.

Henry's Law is sometimes written as: $k' = \frac{C_X}{P_X}$

where k' is also referred to as the Henry's Law constant. As can be seen by comparing equations (1) and (2) above, k' is the reciprocal of k. Since both may be referred to as the Henry's Law constant, readers of the technical literature must be quite careful to note which version of the Henry's Law equation is being used.

Henry's Law is an approximation that only applies for dilute, ideal solutions and for solutions where the liquid solvent does not react chemically with the gas being dissolved.

Partial pressure in diving breathing gases

In recreational diving and professional diving the richness of individual component gases of breathing gases is expressed by partial pressure.

Using diving terms, partial pressure is calculated as:

partial pressure = total absolute pressure x volume fraction of gas component

For the component gas "i":

ppi = P x Fi

For example, at 50 metres (165 feet), the total absolute pressure is 6 bar (600 kPa) (i.e., 1 bar of atmospheric pressure + 5 bar of water pressure) and the partial pressures of the main components of air, oxygen 21% by volume and nitrogen 79% by volume are:

ppN2 = 6 bar x 0.79 = 4.7 bar absolute

ppO2 = 6 bar x 0.21 = 1.3 bar absolute

where:

ppi = partial pressure of gas component i = Pi in the terms used in this article

P = total pressure = P in the terms used in this article

Fi = volume fraction of gas component i = mole fraction, xi, in the terms used in this article

ppN2 = partial pressure of nitrogen = in the terms used in this article

ppO2 = partial pressure of oxygen = in the terms used in this article

The minimum safe lower limit for the partial pressures of oxygen in a gas mixture is 0.16 bar (16 kPa) absolute. Hypoxia and sudden unconsciousness becomes a problem with an oxygen partial pressure of less than 0.16 bar absolute. Oxygen toxicity, involving convulsions, becomes a problem when oxygen partial pressure is too high.

The NOAA Diving Manual recommends a maximum single exposure of 45 minutes at 1.6 bar absolute, of 120 minutes at 1.5 bar absolute, of 150 minutes at 1.4 bar absolute, of 180 minutes at 1.3 bar absolute and of 210 minutes at 1.2 bar absolute. Oxygen toxicity becomes a risk when these oxygen partial pressures and exposures are exceeded. The partial pressure of oxygen determines the maximum operating depth of a gas mixture.

Nitrogen narcosis is a problem when breathing gases at high pressure. Typically, the maximum total partial pressure of narcotic gases used when planning for technical diving is 4.5 bar absolute, based on an equivalent narcotic depth of 35 metres (115 ft).

Shear Stress

A shear stress, denoted (Greek: tau), is defined as a stress which is applied parallel or tangential to a face of a material, as opposed to a normal stress which is applied perpendicularly.

The formula to calculate average shear stress is: $\tau = \frac{F}{A}$,

where

ô = the shear stress;

F = the force applied;

A = the cross sectional area.

Other forms of shear stress

Beam shear

Beam shear is defined as the internal shear stress of a beam caused by the shear force applied to the beam. $\tau = \frac{VQ}{It}$,

where

V = total shear force at the location in question;

Q = statical moment of area;

t = thickness in the material perpendicular to the shear;

I = Moment of Inertia of the entire cross sectional area.

This formula is also known as the Jourawski formula.

Semi-monocoque shear

Shear stresses within a semi-monocoque structure may be calculated by idealizing the cross-section of the structure into a set of stringers (carrying only axial loads) and webs (carrying only shear flows). Dividing the shear flow by the thickness of a given portion of the semi-monocoque structure yields the shear stress. Thus, the maximum shear stress will occur either in the web of maximum shear flow or minimum thickness.

Also constructions in soil can fail due to shear; e.g., the weight of an earth-filled dam or dike may cause the subsoil to collapse, like a small landslide.

Impact shear

The maximum shear stress created in a solid round bar subject to impact is given as the equation: $\tau = 2\left(\frac{UG}{V}\right)^{\frac{1}{2}}$,

where

U = change in kinetic energy;

G = shear modulus;

V = volume of rod;

and $U = U_{rotating} + U_{applied}$; $U_{rotating} = \frac{1}{2}I\omega^2$; $U_{applied} = T\theta_{displaced}$;

I = mass moment of inertia;

w = angular speed.

Shear stress in fluids

Any real fluids (liquids and gases included) moving along solid boundary will incur a shear stress on that boundary. The no-slip condition dictates that the speed of the fluid at the boundary (relative to the boundary) is zero, but at some height from the boundary the flow speed must equal that of the fluid. The region between these two points is aptly named the boundary layer. For all Newtonian fluids in laminar flow the shear stress is proportional to the strain rate in the fluid where the viscosity is the constant of proportionality.

However for Non Newtonian fluids, this is no longer the case as for these fluids the viscosity is not constant. The shear stress is imparted onto the boundary as a result of this loss of velocity. The shear stress, for a Newtonian fluid, at a surface element parallel to a flat plate, at the point y, is given by: $\tau(y) = \mu \frac{\partial u}{\partial y}$,

where

m is the dynamic viscosity of the fluid;

u is the velocity of the fluid along the boundary;

y is the height above the boundary.

Specifically, the wall shear stress is defined as: $\tau_w \equiv \tau(y=0) = \mu \left.\frac{\partial u}{\partial y}\right|_{y=0}$.

In case of wind, the shear stress at the boundary is called wind stress.

MEASUREMENT BY SHEAR STRESS SENSORS

Diverging fringe shear stress sensor

This relationship can be exploited to measure the wall shear stress. If a sensor could directly measure the gradient of the velocity profile at the wall, then multiplying by the dynamic viscosity would yield the shear stress. Such a sensor was demonstrated by A. A. Naqwi and W. C. Reynolds.

The interference pattern generated by sending a beam of light through two parallel slits forms a network of linearly diverging fringes that seem to originate from the plane of the two slits. As a particle in a fluid passes through the fringes, a receiver detects the reflection of the fringe pattern. The signal can be processed, and knowing the fringe angle, the height and velocity of the particle can be extrapolated.

Micro-pillar shear-stress sensor

A further technique recently proposed is that of slender wall-mounted micro-pillars made of the flexible polymer PDMS, which bend in reaction to the applying drag forces in the vicinity of the wall. The deflection of the pillar tips from a reference position is detected optically and serves as a representative of the wall-shear stress. It allows the instantaneous detection of the streamwise and spanwise wall-shear stress distribution in turbulent flow up to high Reynolds numbers.

Pump

A pump is a device used to move fluids, such as liquids, gases or slurries. A pump displaces a volume by physical or mechanical action. Pumps fall into three major groups: direct lift, displacement, and gravity pumps. Their names describe the method for moving a fluid. A positive displacement pump has an expanding cavity on the suction side and a decreasing cavity on the discharge side. Liquid flows into the pump as the cavity on the suction side expands and the liquid flows out of the discharge as the cavity collapses. The volume is constant given each cycle of operation.

A positive displacement pump can be further classified according to the mechanism used to move the fluid:

- Rotary-type, internal gear, screw, shuttle block, flexible vane or sliding vane, circumferential piston, helical twisted roots (e.g. the Wendelkolben pump) or liquid ring vacuum pumps.

Positive displacement rotary pumps are pumps that move fluid using the principles of rotation. The vacuum created by the rotation of the pump captures and draws in the liquid. Rotary pumps are very efficient because they naturally remove air from the lines, eliminating the need to bleed the air from the lines manually.

Positive displacement rotary pumps also have their weaknesses. Because of the nature of the pump, the clearance between the rotating pump and the outer edge must be very close, requiring that the pumps rotate at a slow, steady speed. If rotary pumps are operated at high speeds, the fluids will cause erosion. Rotary pumps that experience such erosion eventually show signs of enlarged clearances, which allow liquid to slip through and detract from the efficiency of the pump.

Positive displacement rotary pumps can be grouped into three main types. Gear pumps are the simplest type of rotary pumps, consisting of two gears laid out side-by-side with their teeth enmeshed. The gears turn away from each other, creating a current that traps fluid between the teeth on the gears and the outer casing, eventually releasing the fluid on the discharge side of the pump as the teeth mesh and go around again. Many small teeth maintain a constant flow of fluid, while fewer, larger teeth create a tendency for the pump to discharge fluids in short, pulsing gushes.

Screw pumps are a more complicated type of rotary pumps, featuring two or three screws with opposing thread -- that is, one screw turns clockwise, and the other counterclockwise. The screws are each mounted on shafts that run parallel to each other; the shafts also have gears on them that mesh with each other in order to turn the shafts together and keep everything in place. The turning of the screws, and consequently the shafts to which they are mounted, draws the fluid through the pump. As with other forms of rotary pumps, the clearance between moving parts and the pump's casing is minimal.

Moving vane pumps are the third type of rotary pumps, consisting of a cylindrical rotor encased in a similarly shaped housing. As the rotor turns, the vanes trap fluid between the rotor and the casing, drawing the fluid through the pump.

- Reciprocating-type, for example, piston or diaphragm pumps.

Positive displacement pumps have an expanding cavity on the suction side and a decreasing cavity on the discharge side. Liquid flows into the pumps as the cavity on the suction side expands and the liquid flows out of the discharge as the cavity collapses. The volume is constant given each cycle of operation.

The positive displacement pumps can be divided into two main classes:

- Reciprocating
- Rotary

The positive displacement principle applies whether the pump is a:

- Rotary lobe pump
- Progressive cavity pump
- Rotary gear pump
- Piston pump
- Diaphragm pump
- Screw pump
- Gear pump
- Hydraulic pump
- Vane pump
- Regenerative (peripheral) pump
- Peristaltic pump

Positive displacement pumps, unlike centrifugal or roto-dynamic pumps, will produce the same flow at a given speed (RPM) no matter what the discharge pressure.

- Positive displacement pumps are "constant flow machines"

A positive displacement pump must not be operated against a closed valve on the discharge side of the pump because it has no shut-off head like centrifugal pumps. A positive displacement pump operating against a closed discharge valve, will continue to produce flow until the pressure in the discharge line are increased until the line bursts or the pump is severely damaged - or both.

A relief or safety valve on the discharge side of the positive displacement pump is therefore necessary. The relief valve can be internal or external. The pump manufacturer normally has the option to supply internal relief or safety valves. The internal valve should in general only be used as a safety precaution, an external relief valve installed in the discharge line with a return line back to the suction line or supply tank is recommended.

Reciprocating pumps

Typical reciprocating pumps are

- Plunger pumps
- Diaphragm pumps

A plunger pump consists of a cylinder with a reciprocating plunger in it. The suction and discharge valves are mounted in the head of the cylinder. In the suction stroke the plunger retracts and the suction valves open causing suction of fluid into the cylinder. In the forward stroke the plunger pushes the liquid out of the discharge valve.

With only one cylinder the fluid flow varies between maximum flow when the plunger moves through the middle positions, and zero flow when the plunger is at the end positions. A lot of energy is wasted when the fluid is accelerated in the piping system. Vibration and "water hammer" may be a serious problem. In general the problems are compensated for by using two or more cylinders not working in phase with each other.

In diaphragm pumps, the plunger pressurizes hydraulic oil which is used to flex a diaphragm in the pumping cylinder. Diaphragm valves are used to pump hazardous and toxic fluids.

An example of the piston displacement pump is the common hand soap pump.

Gear pump

This uses two meshed gears rotating in a closely fitted casing. Fluid is pumped around the outer periphery by being trapped in the tooth spaces. It does not travel back on the meshed part, since the teeth mesh closely in the centre. Widely used on car engine oil pumps. it is also used in various hydraulic power packs..

Progressing cavity pump

Widely used for pumping difficult materials such as sewage sludge contaminated with large particles, this pump consists of a helical shaped rotor, about ten times as long as its width. This can be visualized as a central core of diameter x, with typically a curved spiral wound around of thickness half x, although of course in reality it is made from one casting. This shaft fits inside a heavy duty rubber sleeve, of wall thickness typically x also. As the shaft rotates, fluid is gradually forced up the rubber sleeve. Such pumps can develop very high pressure at quite low volumes.

Roots-type pumps

The low pulsation rate and gentle performance of this Roots-type positive displacement pump is achieved due to a combination of its two 90° helical twisted rotors, and a triangular shaped sealing line configuration, both at the point of suction and at the point of discharge. This design produces a continuous and non-vorticuless flow with equal volume. High capacity industrial "air compressors" have been designed to employ this principle, as well as most "superchargers" used on internal combustion engines, and even a brand of civil defense siren, the Federal Signal Corporation's Thunderbolt.

Peristaltic pump

A peristaltic pump is a type of positive displacement pump used for pumping a variety of fluids. The fluid is contained within a flexible tube fitted inside a circular pump casing (though linear peristaltic pumps have been made). A rotor with a number of "rollers", "shoes" or "wipers" attached to the external circumference compresses the flexible tube.

As the rotor turns, the part of the tube under compression closes (or "occludes") thus forcing the fluid to be pumped to move through the tube. Additionally, as the tube opens to its natural state after the passing of the cam ("restitution") fluid flow is induced to the pump. This process is called peristalsis and is used in many biological systems such as the gastrointestinal tract.

Reciprocating-type pumps

Reciprocating pumps are those which cause the fluid to move using one or more oscillating pistons, plungers or membranes (diaphragms). Reciprocating-type pumps require a system of suction and discharge valves to ensure that the fluid moves in a positive direction. Pumps in this category range from having "simplex" one cylinder, to in some cases "quad" four cylinders or more. Most reciprocating-type pumps are "duplex" (two) or "triplex" (three) cylinder.

Furthermore, they can be either "single acting" independent suction and discharge strokes or "double acting" suction and discharge in both directions. The pumps can be powered by air, steam or through a belt drive from an engine or motor. This type of pump was used extensively in the early days of steam propulsion (19th century) as boiler feed water pumps. Reciprocating pumps are now typically used for pumping highly viscous fluids including concrete and heavy oils, and special applications demanding low flow rates against high resistance.

Compressed-air-powered double-diaphragm pumps

One modern application of positive displacement diaphragm pumps is compressed-air-powered double-diaphragm pumps. Run on compressed air these pumps are intrinsically safe by design, although all manufacturers offer ATEX certified models to comply with industry regulation. Commonly seen in all areas of industry from shipping to processing, SandPiper, Wilden Pumps or ARO are generally the larger of the brands.

They are relatively inexpensive and can be used for almost any duty from pumping water out of bunds, to pumping hydrochloric acid from secure storage (dependent on how the pump is manufactured - elastomers / body construction). Lift is normally limited to roughly 6m although heads can reach almost 200 Psi. A hydraulic ram is a water pump powered by hydropower.

It functions as a hydraulic transformer that takes in water at one "hydraulic head" (pressure) and flow-rate, and outputs water at a higher hydraulic-head and lower flow-rate. The device utilizes the water hammer effect to develop pressure that allows a portion of the input water that powers the pump to be lifted to a point higher than where the water originally started.

The hydraulic ram is sometimes used in remote areas, where there is both a source of low-head hydropower, and a need for pumping water to a destination higher in elevation than the source. In this situation, the ram is often useful, since it requires no outside source of power other than the kinetic energy of flowing water..

Velocity pumps

Rotodynamic pumps (or dynamic pumps) are a type of velocity pump in which kinetic energy is added to the fluid by increasing the flow velocity. This increase in energy is converted to a gain in potential energy (pressure) when the velocity is reduced prior to or as the flow exits the pump into the discharge pipe. This conversion of kinetic energy to pressure can be explained by the First law of thermodynamics or more specifically by Bernoulli's principle. Dynamic pumps can be further subdivided according to the means in which the velocity gain is achieved.

These types of pumps have a number of characteristics:

1. Continuous energy
2. Conversion of added energy to increase in kinetic energy (increase in velocity)
3. Conversion of increased velocity (kinetic energy) to an increase in pressure head

One practical difference between dynamic and positive displacement pumps is their ability to operate under closed valve conditions. Positive displacement pumps physically displace the fluid; hence closing a valve downstream of a positive displacement pump will result in a continual build up in pressure resulting in mechanical failure of either pipeline or pump. Dynamic pumps differ in that they can be safely operated under closed valve conditions (for short periods of time).

Centrifugal pump

A centrifugal pump is a rotodynamic pump that uses a rotating impeller to increase the pressure and flow rate of a fluid. Centrifugal pumps are the most common type of pump used to move liquids through a piping system. The fluid enters the pump impeller along or near to the rotating axis and is accelerated by the impeller, flowing radially outward or axially into a diffuser or volute chamber, from where it exits into the downstream piping system. Centrifugal pumps are typically used for large discharge through smaller heads.

Centrifugal pumps are most often associated with the radial flow type. However, the term "centrifugal pump" can be used to describe all impeller type rotodynamic pumps including the radial, axial and mixed flow variations.

Radial flow pumps

Often simply referred to as centrifugal pumps. The fluid enters along the axial plane, is accelerated by the impeller and exits at right angles to the shaft (radially). Radial flow pumps operate at higher pressures and lower flow rates than axial and mixed flow pumps.

Axial flow pumps

Axial flow pumps differ from radial flow in that the fluid enters and exits along the same direction parallel to the rotating shaft. The fluid is not accelerated but instead "lifted" by the action of the impeller. They may be likened to a propeller spinning in a length of tube. Axial flow pumps operate at much lower pressures and higher flow rates than radial flow pumps.

Mixed flow pumps

Mixed flow pumps, as the name suggests, function as a compromise between radial and axial flow pumps, the fluid experiences both radial acceleration and lift and exits the impeller somewhere between 0-90 degrees from the axial direction. As a consequence mixed flow pumps operate at higher pressures than axial flow pumps while delivering higher discharges than radial flow pumps. The exit angle of the flow dictates the pressure head-discharge characteristic in relation to radial and mixed flow.

Eductor-jet pump

This uses a jet, often of steam, to create a low pressure. This low pressure sucks in fluid and propels it into a higher pressure region.

Gravity pumps

Gravity pumps include the syphon and Heron's fountain - and there also important qanat or foggara systems which simply use downhill flow to take water from far-underground aquifers in high areas to consumers at lower elevations. The hydraulic ram is also sometimes referred to as a gravity pump.

Steam pumps

Steam pumps are now mainly of historical interest. They include any type of pump powered by a steam engine and also pistonless pumps such as Thomas Savery's pump and the Pulsometer steam pump.

Valveless pumps

Valveless pumping assists in fluid transport in various biomedical and engineering systems. In a valveless pumping system, no valves are present to regulate the flow direction.

The fuid pumping efficiency of a valveless system, however, is not necessarily lower than that having valves.

In fact, many fluid-dynamical systems in nature and engineering more or less rely upon valveless pumping to transport the working fluids therein.

For instance, blood circulation in the cardiovascular system is maintained to some extent even when the heart's valves fail. Meanwhile, the embryonic vertebrate heart begins pumping blood long before the development of discernable chambers and valves. In microfuidics, valveless impedance pump have been fabricated, and are expected to be particularly suitable for handling sensitive biofuids.

PUMP REPAIRS

Examining pump repair records and MTBF (mean time between failures) is of great importance to responsible and conscientious pump users. In view of that fact, the preface to the 2006 Pump User's Handbook alludes to "pump failure" statistics. For the sake of convenience, these failure statistics often are translated into MTBF (in this case, installed life before failure).

In early 2005, Gordon Buck, John Crane Inc.'s chief engineer for Field Operations in Baton Rouge, LA, examined the repair records for a number of refinery and chemical plants to obtain meaningful reliability data for centrifugal pumps. A total of 15 operating plants having nearly 15,000 pumps were included in the survey. The smallest of these plants had about 100 pumps; several plants had over 2000. All facilities were located in the United States. In addition, considered as "new," others as "renewed" and still others as "established." Many of these plants-but not all-had an alliance arrangement with John Crane. In some cases, the alliance contract included having a John Crane Inc. technician or engineer on-site to coordinate various aspects of the program.

Not all plants are refineries, however, and different results can be expected elsewhere. In chemical plants, pumps have traditionally been "throw-away" items as chemical attack can result in limited life. Things have improved in recent years, but the somewhat restricted space available in "old" DIN and ASME-standardized stuffing boxes places limits on the type of seal that can be fitted. Unless the pump user upgrades the seal chamber, only the more compact and simple versions can be accommodated. Without this upgrading, lifetimes in chemical installations are generally believed to be around 50 to 60 percent of the refinery values.

It goes without saying that unscheduled maintenance often is one of the most significant costs of ownership, and failures of mechanical seals and bearings are among the major causes. Keep in mind the potential value of selecting pumps that cost more initially, but last much longer between repairs. The MTBF of a better pump may be one to four years longer than that of its non-upgraded counterpart. Consider that published average values of avoided pump failures range from $2600 to $12,000. This does not include lost opportunity costs. One pump fire occurs per 1000 failures. Having fewer pump failures means having fewer destructive pump fires.

As has been noted, a typical pump failure based on actual year 2002 reports, costs $5,000 on average. This includes costs for material, parts, labor and overhead. Let us now assume that the MTBF for a particular pump is 12 months and that it could be extended to 18 months. This would result in a cost avoidance of $2,500/ yr-which is greater than the premium one would pay for the reliability-upgraded centrifugal pump.

Applications

Pumps are used throughout society for a variety of purposes. Early applications includes the use of the windmill or watermill to pump water. Today, the pump is used for irrigation, water supply, gasoline supply, air conditioning systems, refrigeration (usually called a compressor), chemical movement, sewage movement, flood control, marine services, etc.

Because of the wide variety of applications, pumps have a plethora of shapes and sizes: from very large to very small, from handling gas to handling liquid, from high pressure to low pressure, and from high volume to low volume.

Priming a pump

Liquid and slurry pumps can lose prime and this will require the pump to be primed by adding liquid to the pump and inlet pipes to get the pump started. Loss of "prime" is usually due to ingestion of air into the pump. The clearances and displacement ratios in pumps used for liquids and other more viscous fluids cannot displace the air due to its lower density.

Pumps as public water supplies

One sort of pump once common worldwide was a hand-powered water pump, or 'pitcher pump'. It would be installed over a community water well that was used by people in the days before piped water supplies.

In parts of the British Isles, it was often called "the parish pump". Although such community pumps are no longer common, the expression "parish pump" is still used. It derives from the kind of the chatter and conversation that might be heard as people congregated to draw water from the community water pump, and is now used to describe a place or forum where matter of purely local interest is discussed.

Because water from pitcher pumps is drawn directly from the soil, it is more prone to contamination. If such water is not filtered and purified, consumption of it might lead to gastrointestinal or other water-borne diseases.

Modern hand operated community pumps are considered the most sustainable low cost option for safe water supply in resource poor settings, often in rural areas in developing countries. A hand pump opens access to deeper groundwater that is often not polluted and also improves the safety of a well by protecting the water source from contaminated buckets. Pumps like the Afridev pump are designed to be cheap to build and install, and easy to maintain with simple parts. However, scarcity of spare parts for these type of pumps in some regions of Africa has diminished their utility for these areas.

SEALING MULTIPHASE PUMPING APPLICATIONS

Multiphase pumping applications, also referred to as tri-phase, have grown due to increased oil drilling activity. In addition, the economics of multiphase production is attractive to upstream operations as it leads to simpler, smaller in-field installations, reduced equipment costs and improved production rates. In essence, the multiphase pump can accommodate all fluid stream properties with one piece of equipment, which has a smaller footprint. Often, two smaller multiphase pumps are installed in series rather than having just one massive pump.

For midstream and upstream operations, multiphase pumps can be located onshore or offshore and can be connected to single or multiple wellheads. Basically, multiphase pumps are used to transport the untreated flow stream produced from oil wells to downstream processes or gathering facilities. This means that the pump may handle a flow stream (well stream) from 100 percent gas to 100 percent liquid and every imaginable combination in between. The flow stream can also contain abrasives such as sand and dirt. Multiphase pumps are designed to operate under changing/fluctuating process conditions. Multiphase pumping also helps eliminate emissions of greenhouse gases as operators strive to minimize the flaring of gas and the venting of tanks where possible.

Types and Features of Multiphase Pumps

Helico-Axial Pumps (Centrifugal) A rotodynamic pump with one single shaft requiring two mechanical seals. This pump utilizes an open-type axial impeller. This pump type is often referred to as a "Poseidon Pump" and can be described as a cross between an axial compressor and a centrifugal pump.

Twin Screw (Positive Displacement) The twin screw pump is constructed of two intermeshing screws that force the movement of the pumped fluid. Twin screw pumps are often used when pumping conditions contain high gas volume fractions and fluctuating inlet conditions. Four mechanical seals are required to seal the two shafts.

Progressive Cavity Pumps (Positive Displacement) Progressive cavity pumps are single-screw types typically used in shallow wells or at the surface. This pump is mainly used on surface applications where the pumped fluid may contain a considerable amount of solids such as sand and dirt.

Electric Submersible Pumps (Centrifugal) These pumps are basically multistage centrifugal pumps and are widely used in oil well applications as a method for artificial lift. These pumps are usually specified when the pumped fluid is mainly liquid.

Buffer Tank A buffer tank is often installed upstream of the pump suction nozzle in case of a slug flow. The buffer tank breaks the energy of the liquid slug, smoothes any fluctuations in the incoming flow and acts as a sand trap.

As the name indicates, multiphase pumps and their mechanical seals can encounter a large variation in service conditions such as changing process fluid composition, temperature variations, high and low operating pressures and exposure to abrasive/erosive media. The challenge is selecting the appropriate mechanical seal arrangement and support system to ensure maximized seal life and its overall effectiveness.

Specifications

Pumps are commonly rated by horsepower, flow rate, outlet pressure in feet (or metres) of head, inlet suction in suction feet (or metres) of head. The head can be simplified as the number of feet or metres the pump can raise or lower a column of water at atmospheric pressure.

From an initial design point of view, engineers often use a quantity termed the specific speed to identify the most suitable pump type for a particular combination of flow rate and head.

Pump material

Pump material can be of Stainless steel (SS 316 or SS 304), cast iron etc. It depend upon the application of pump. In water industry for pharma application, SS 316 is normally used. As at high temperature stainless steel give better result.

Pumping power

The power imparted into a fluid will increase the energy of the fluid per unit volume. Thus the power relationship is between the conversion of the mechanical energy of the pump mechanism and the fluid elements within the pump. In general, this is governed by a series of simultaneous differential equations, known as the Navier-Stokes equations. However a more simple equation relating only the different energies in the fluid, known as Bernoulli's equation can be used. Hence the power, P, required by the pump:

where DP is the change in total pressure between the inlet and outlet (in Pa), and Q, the fluid flowrate is given in m^3/s. The total pressure may have gravitational, static pressure and kinetic energy components; i.e. energy is distributed between change in the fluid's gravitational potential energy (going up or down hill), change in velocity, or change in static pressure. h is the pump efficiency, and may be given by the manufacturer's information, such as in the form of a pump curve, and is typically derived from either fluid dynamics simulation (i.e. solutions to the Navier-stokes for the particular pump geometry), or by testing. The efficiency of the pump will depend upon the pump's configuration and operating conditions (such as rotational speed, fluid density and viscosity etc).

$$\Delta P = \frac{(v_2^2 - v_1^2)}{2} + \Delta z g + \frac{\Delta p_{\mathrm{static}}}{\rho}$$

For a typical "pumping" configuration, the work is imparted on the fluid, and is thus positive. For the fluid imparting the work on the pump (i.e. a turbine), the work is negative power required to drive the pump is determined by dividing the output power by the pump efficiency. Furthermore, this definition encompasses pumps with no moving parts, such as a siphon.

Pump efficiency

Pump efficiency is defined as the ratio of the power imparted on the fluid by the pump in relation to the power supplied to drive the pump. Its value is not fixed for a given pump, efficiency is a function of the discharge and therefore also operating head. For centrifugal pumps, the efficiency tends to increase with flow rate up to a point midway through the operating range (peak efficiency) and then declines as flow rates rise further. Pump performance data such as this is usually supplied by the manufacturer before pump selection. Pump efficiencies tend to decline over time due to wear (e.g. increasing clearances as impellers reduce in size).

One important part of system design involves matching the pipeline headloss-flow characteristic with the appropriate pump or pumps which will operate at or close to the point of maximum efficiency. There are free tools that help calculate head needed and show pump curves including their Best Efficiency Points (BEP).

Pump efficiency is an important aspect and pumps should be regularly tested. Thermodynamic pump testing is one method.

Pump selection is done by performance curve which is curve between pressure head and flow rate. And also power supply is also taken care of. Pumps are normally available that run at 50 hz or 60 hz.

9

MATERIAL HANDLING

Material handling equipment is all equipment that relates to the movement, storage, control and protection of materials, goods and products throughout the process of manufacturing, distribution, consumption and disposal. Material handling equipment is the mechanical equipment involved in the complete system. Material handling equipment is generally separated into four main categories: storage and handling equipment, engineered systems, industrial trucks, and bulk material handling.

Material handling equipment is used to increase throughput, control costs, and maximize productivity. There are several ways to determine if the material handling equipment is achieving peak efficiency. These include capturing all relevant data related to the warehouse's operation (such as SKUs), measuring how many times an item is "touched" from the time it is ordered until it leaves the building, making sure you are using the proper picking technology, and keeping system downtime to a minimum.

Bulk material handling is an engineering field that is centered around the design of equipment used for the transportation of materials such as ores and cereals in loose bulk form. It can also relate to the handling of mixed wastes.

Bulk material handling systems are typically composed of moveable items of machinery such as conveyor belts, stackers, reclaimers, bucket elevators, shiploaders, unloaders and various shuttles hoppers and diverters combined with storage facilities such as stockyards, storage silos or stockpiles.

The purpose of a bulk material handling facility is generally to transport material from one of several locations (i.e. a source) to an ultimate destination. Providing storage and inventory control and possibly material blending is usually part of a bulk material handling system.

Bulk material handling systems can be found on mine sites, ports (for loading or unloading of cereals, ores and minerals) and processing facilities (such as iron and steel, coal fired power stations refineries).

In ports handling large quantities of bulk materials continuous ship unloaders are replacing gantry cranes.

Types of material handling equipment

Storage and handling equipment

Storage and handling equipment is a category within the material handling industry. The equipment that falls under this description is usually non-automated storage equipment. Products such as Pallet rack, shelving, carts, etc. belong to storage and handling. Many of these products are often referred to as "catalog" items because they generally have globally accepted standards and are often sold as stock materials out of Material handling catalogs.

Engineered systems

Engineered systems are typically custom engineered material handling systems. Conveyors, Handling Robots, AS/RS, AGV and most other automated material handling systems fall into this category. Engineered systems are often a combination of products integrated to one system. Many distribution centers will optimize storage and picking by utilizing engineered systems such as pick modules and sortation systems.

Equipment and utensils used for processing or otherwise handling edible product or ingredients must be of such material and construction to facilitate thorough cleaning and to ensure that their use will not cause the adulteration of product during processing, handling, or storage. Equipment and utensils must be maintained in sanitary condition so as not to adulterate product.

Industrial trucks

Industrial trucks usually refer to operator driven motorized warehouse vehicles, powered manually, by gasoline, propane or electrically. Industrial trucks assist the material handling system with versatility; they can go where engineered systems cannot. Forklift trucks are the most common example of industrial trucks but certainly aren't the extent of the category. Tow tractors and stock chasers are additional examples of industrial trucks.

Bulk material handling

Bulk material handling equipment is used to move and store bulk materials such as ore, liquids, and cereals. This equipment is often seen on farms, mines, shipyards and refineries.

SILO

A silo is a structure for storing bulk materials. Silos are used in agriculture to store grain or fermented feed known as silage. Silos are more commonly used for bulk storage of grain, coal, cement, carbon black, woodchips, food products and sawdust. Three types of silos are in widespread use today - tower silos, bunker silos and bag silos. Missile silos are used for the storage and launching of ballistic missiles. There are different types of cement silos such as the low-level mobile silo and the static upright cement silo, which are used to hold and discharge cement and other powder materials such as PFA (Pulverised Fuel Ash). The low-level silos are fully mobile with capacities from 10 to 75 tons. They are simple to transport and are easy to set up on site. These mobile silos generally come equipped with an electronic weighing system with digital display and printer.

This allows any quantity of cement or powder discharged from the silo to be controlled and also provides an accurate indication of what remains inside the silo. The static upright silos have capacities from 20 to 80 tons. These are considered a low-maintenance option for the storage of cement or other powders. Cement silos can be used in conjunction with bin-fed batching plants.

Tower silo

Storage silos are cylindrical structures, typically 10 to 90 ft (4 to 30 m) in diameter and 30 to 275 ft (10 to 84 m) in height with the slipform and Jumpform concrete silos being the larger diameter and taller silos. They can be made of many materials. Wood staves, concrete staves, cast concrete, and steel panels have all been used, and have varying cost, durability, and airtightness tradeoffs. Silos storing grain, cement and woodchips are typically unloaded with air slides or augers. Silos can be unloaded into rail cars, trucks or conveyors.

Tower silos containing silage are usually unloaded from the top of the pile, originally by hand using a silage fork, which has many more tines than the common pitchfork, 12 vs 4, in modern times using mechanical unloaders. Bottom silo unloaders are utilized at times but have problems with difficulty of repair.

An advantage of tower silos is that the silage tends to pack well due to its own weight, except in the top few feet. However, this may be a disadvantage for items like chopped wood. The tower silo was invented by Franklin Hiram King.

In Australia and the United States, many country towns or the larger farmers in grain-growing areas have groups of concrete tower silos, known as grain elevators, to collect grain from the surrounding towns and store and protect the grain for transport by train, truck or barge to a processor or to an export port. The transport system simply can not handle the peak load due of the harvest needs which depend upon the weather. In bumper crop times, the excess grain is stored in piles without silos or bins, causing considerable looses.

Concrete stave silos are constructed from small precast concrete blocks with ridged grooves along each edge that lock them together into a high strength shell. Much of concrete's strength comes from its high compressibility, so the silo is held together by steel hoops encircling the tower and compressing the staves into a tight ring. The vertical stacks are held together by intermeshing of the ends of the staves by a short distance around the perimeter of each layer, and hoops which are tightened directly across the stave edges.

The static pressure of the material inside the silo pressing outward on the staves increases towards the bottom of the silo, so the hoops can be spaced wide apart near the top but become progressively more closely spaced towards the bottom to prevent seams from opening and the contents leaking out.

Concrete stave silos are built from common components designed with high strength and long life. They have the flexibility to have their height increased according to the needs of the farm and purchasing power of the farmer, or to be completely disassembled and reinstalled somewhere else if no longer needed.

Low-oxygen tower silos

Low-oxygen silos (commonly seen as blue Harvestore farm silos) are designed to keep the contents in a low-oxygen atmosphere at all times to keep the fermented contents in a high quality state and to prevent mold and decay, as may occur in the top layers of a stave silo or bunker. These silos are only opened directly to the atmosphere during the initial forage loading, and even the unloader chute is sealed.

It would be expensive to design a such a large structure that is immune to atmospheric pressure changes over time. Instead, the silo structure is open to the atmosphere but outside air is separated from internal air by large impermeable bags sealed to the silo breather openings. In the warmth of the day when the silo is heated by the sun, the gas trapped inside the silo expands and the bags "breathe out" and collapse. At night the silo cools, the air inside contracts and the bags "breathe in" and expand again.

While the iconic blue Harvestore low-oxygen silos were once very common, the speed of its unloader mechanism was not able to match the output rates of modern bunker silos and this type of silo went into decline. Unloader repair expenses also severely hurt the Harvestore reputation, because the unloader feed mechanism is located in the bottom of the silo under tons of silage. In the event of cutter chain breakage, it can cost up to US$10,000 to perform repairs. The silo may need to be partially or completely emptied by alternate means, to unbury the broken unloader retrieve broken components lost in the silage at the bottom of the structure.

In 2005 the Harvestore company recognized these issues and worked to develop new unloaders with double the flow rate of previous models to stay competitive with bunkers, and with far greater unloader chain strength. They are now also using load sensing soft-start variable frequency drive motor controllers to reduce the likelihood of mechanism breakage, and to control the feeder sweep arm movement.

While the sight of multiple tall silos may look impressive, many have been abandoned for silage use, or converted to hold grain. Ground-level bunker silos carry none of the hazards outlined above, and need only a standard tractor/loader to feed out with, and no specialized machinery.

Bunker silos

Bunker silos are trenches, usually with concrete walls, that are filled and packed with tractors and loaders. The filled trench is covered with a plastic tarp to make it airtight. These silos are usually unloaded with a tractor and loader. They are inexpensive and especially well suited to very large operations.

Bag silos

Bag silos are heavyweight plastic tubes, usually around 8 to 12 ft in diameter, and of variable length as required for the amount of material to be stored. They are packed using a machine made for the purpose, and sealed on both ends. They are unloaded using a tractor and loader or skid-steer loader. The bag is discarded in sections as it is torn off. Bag silos require little capital investment. They can be used as a temporary measure when growth or harvest conditions require more space, though some farms use them every year.

Bins

A bin is typically much shorter than a silo, and is typically used for holding dry matter such as concrete or grain. Bins may be round or square, but round bins tend to empty more easily due to a lack of corners for the stored material to become wedged and encrusted.

The stored material may be powdered, as seed kernels, or as cob corn. Due to the dry nature of the stored material, it tends to be lighter than silage and can be more easily handled by under-floor grain unloaders. To facilitate drying after harvesting, some grain bins contain a hollow perforated or screened central shaft to permit easier air infiltration into the stored grain.

Sand and Salt Silos

Sand and salt for winter road maintenance are stored in conical dome shaped silos. These are more common in North America, namely in Canada and the United States.

Forage Silo usage

Forage silo filling is performed using a forage harvester which may either be self-propelled with an engine and driver's cab, or towed behind a tractor that supplies power through a PTO.

The harvester contains a drum-shaped series of cutting knives which shear the fibrous plant material into small pieces no more than an inch long, to facilitate mechanized blowing and transport via augers. The finely chopped plant material is then blown by the harvester into a forage wagon which contains an automatic unloading system.

Tower forage filling is typically performed with a silo blower which is a very large fan with paddle-shaped blades. Material is fed into a vibrating hopper and is pushed into the blower using a spinning spiral auger.

There is commonly a water connection on the blower to add moisture to the plant matter being blown into the silo. The blower may be driven by an electric motor but it is more common to use a spare tractor instead.

A large slow-moving conveyor chain underneath the silage in the forage wagon moves the pile towards the front, where rows of rotating teeth break up the pile and drop it onto a high-speed transverse conveyor that pours the silage out the side of the wagon into the blower hopper.

Bag filling

Silo bags are filled using a traveling sled driven from the PTO of a tractor left in neutral and which is gradually pushed forward as the bag is filled. The steering of the tractor controls the direction of bag placement as it fills, but bags are normally laid in a straight line.

The bag is loaded using the same forage harvesting methods as the tower, but the forage wagon must be moved progressively forward with the bag loader. The loader uses an array of rotating cam-shaped spiraled teeth associated with a large comb-shaped tines to push forage into the bag. The forage is pushed in through a large opening, and as the teeth rotate back out, they pass between the comb tines. The cam-shaped auger teeth essentially wipe the forage off using the steel tines, keeping the forage in the bag.

Before filling begins, the entire bag is placed onto the loader as a bunched-up tube folded back on itself in many layers to form a thick pile of plastic. Because the plastic is minimally elastic, the loader mechanism filling chute is slightly smaller than the final size of the bag, to accommodate this stack of plastic around the mouth of the loader. The plastic slowly unfurls itself around the edges of the loader as the tube is filled.

The contents of the silo bag are under pressure as it is filled, with the pressure controlled by a large brake shoe pressure regulator, holding back two large winch drums on either side of the loader. Cables from the drum extend to the rear of the bag where a large mesh basket holds the rear end of the bag shut.

To prevent molding and to assure an airtight seal during fermentation, the ends of the silo bag tube are gathered, folded, and tied shut to prevent oxygen from entering the bag. Removal of the bag loader can be hazardous to bystanders since the pressure must be released and the rear end allowed to collapse onto the ground.

The main operating component of the silo unloader is suspended in the silo from a steel cable on a pulley that is mounted in the top-center of the roof of the silo. The vertical positioning of the unloader is controlled by an electric winch on the exterior of the silo.

For the summer filling of a tower silo, the unloader is winched as high as possible to the top of the silo and put into a parking position. The silo is filled with a silo blower, which is literally a very large fan that blows a large volume of pressurized air up a 10-inch tube on the side of the silo. A small amount of water is introduced into the air stream during filling to help lubricate the filling tube.

A small adjustable nozzle at the top, controlled by a handle at the base of the silo directs the silage to fall into the silo on the near, middle, or far side, to facilitate evenly layered loading. Once completely filled, the top of the exposed silage pile is covered with a large heavy sheet of silo plastic which seals out oxygen and permits the entire pile to begin to ferment in the autumn.

In the winter when animals must be kept indoors, the silo plastic is removed, the unloader is lowered down onto the top of the silage pile, and a hinged door is opened on the side of the silo to permit the silage to be blown out. There is an array of these access doors arranged vertically up the side of the silo, with an unloading tube next to the doors that has a series of removable covers down the side of the tube. The unloader tube and access doors are normally covered with a large U-shaped shield mounted on the silo, to protect the farmer from wind, snow, and rain while working on the silo.

The silo unloader mechanism consists of a pair of counter-rotating toothed augers which rip up the surface of the silage and pull it towards the center of the unloader. The toothed augers rotate in a circle around the center hub, evenly chewing the silage off the surface of the pile. In the center, a large blower assembly picks up the silage and blows it out the silo door, where the silage falls by gravity down the unloader tube to the bottom of the silo, typically into an automated conveyor system.

The unloader is typically lowered only a half-inch or so at a time by the operator, and the unloader picks up only a small amount of material until the winch cable has become taut and the unloader is not picking up any more material. The operator then lowers the unloader another half-inch or so and the process repeats. If lowered too far, the unloader can pull up much more material than it can handle, which can overflow and plug up the blower, outlet spout, and the unloader tube, resulting in a time-wasting process of having to climb up the silo to clear the blockages.

Once silage has entered the conveyor system, it can be handled by either manual or automatic distribution systems. The simplest manual distribution system uses a sliding metal platform under the pickup channel. When slid open, the forage drops through the open hole and down a chute into a wagon, wheelbarrow, or open pile. When closed, the forage continues past the opening and onward to other parts of the conveyor. Computer automation and a conveyor running the length of a feeding stall can permit the silage to be automatically dropped from above by each animal, with the amount dispensed customized for each location.

TOWER FORAGE SILO HAZARDS

There are also many opportunities for a farmer to be injured, maimed, or killed by the various mechanical systems of the silo loading and unloading process. Filling a silo requires parking two tractors very close to each other, both running at full power and with live PTO shafts, one powering the silo blower and the other powering a forage wagon unloading fresh-cut forage into the blower. The farmer must continually move around in this highly hazardous environment of spinning shafts and high-speed conveyors to check material flows and adjust speeds, and to start and stop all the equipment between loads.

Preparation for filling a silo requires winching the unloader to the top, and any remaining forage at the base that the unloader could not pick up must be removed from the floor of the silo. This job requires that the farmer work directly underneath a machine weighing several tons suspended fifty feet or more overhead from a small steel cable. If the unloader were to fall, the farmer would likely be killed instantly.

Unloading also poses its own special hazards, due to the requirement that the farmer regularly climb the silo to close an upper door and open a lower door, moving the unloader chute from door to door in the process. The fermentation of the silage produces methane gas which over time will outgas and displace the oxygen in the top of the silo.

A farmer directly entering a silo without any other precautions can be asphyxiated by the methane, knocked unconscious, and silently suffocate to death before anyone else knows what has happened. It is either necessary to leave the silo blower attached to the silo at all times to use it when necessary to ventilate the silo with fresh air, or to have a dedicated electric fan system to blow fresh air into the silo, before anyone attempts to enter it.

In the event that the unloader mechanism becomes plugged, the farmer must climb the silo and directly stand on the unloader, reaching into the blower spout to dig out the soft silage. After clearing a plug, the forage needs to be forked out into an even layer around the unloader so that the unloader does not immediately dig into the pile and plug itself again. All during this process the farmer is standing on or near a machine that could easily kill them in seconds if it were to accidentally start up. This can happen if someone in the barn were to unknowingly switch on the unloading mechanism while someone is in the silo working on the unloader.

Dry-material / bin hazards

Silos are hazardous, and people die every year in the process of filling and maintaining them. The machinery used is dangerous and with tower silos workers can fall from the silo's ladder or work platform. Several fires have occurred over the years.

There have also been many cases of silos and the associated ducts and buildings exploding. If the air inside becomes laden with finely granulated particles, such as grain dust, a spark can trigger an explosion powerful enough to blow a concrete silo and adjacent buildings apart, usually setting the adjacent grain and building on fire. Sparks are often caused by (metal) rubbing against metal ducts; or due to static electricity produced by dust moving along the ducts when extra dry.

The two main problems which will necessitate cleaning in dry-matter silos and bins are Bridging and Rat-holing. Bridging occurs when the material interlaces over the unloading mechanism at the base of the silo and blocks the flow of stored material by gravity into the unloading system. Rat-holing occurs when the material starts to adhere to the side of the silo.

This will reduce the operating capacity of a silo as well as leading to cross-contamination of newer material with older material. There are a number of ways to clean a silo and many of these carry their own risks. However since the early 1990s acoustic cleaners have become available. These are non-invasive, have minimum risk, and can offer a very cost-effective way to keep a small particle silo clean.

Conveyor System

A conveyor system is a common piece of mechanical handling equipment that moves materials from one location to another. Conveyors are especially useful in applications involving the transportation of heavy or bulky materials. Conveyor systems allow quick and efficient transportation for a wide variety of materials, which make them very popular in the material handling and packaging industries. Many kinds of conveying systems are available, and are used according to the various needs of different industries.

Conveyor systems are used widespread across a range of industries due to the numerous benefits they provide.

- Conveyors are able to safely transport materials from one level to another, which when done by human labor would be strenuous and expensive.
- They can be installed almost anywhere, and are much safer than using a forklift or other machine to move materials.
- They can move loads of all shapes, sizes and weights. Also, many have advanced safety features that help prevent accidents.
- There are a variety of options available for running conveying systems, including the hydraulic, mechanical and fully automated systems, which are equipped to fit individual needs.

Conveyor systems are commonly used in many industries, including the automotive, agricultural, computer, electronic, food processing, aerospace, pharmaceutical, chemical, bottling and canning, print finishing and packaging. Although a wide variety of materials can be conveyed, some of the most common include food items such as beans and nuts, bottles and cans, automotive components, scrap metal, pills and powders, wood and furniture and grain and animal feed. Many factors are important in the accurate selection of a conveyor system. It is important to know how the conveyor system will be used beforehand. Some individual areas that are helpful to consider are the required conveyor operations, such as transportation, accumulation and sorting, the material sizes, weights and shapes and where the loading and pickup points need to be.

CARE AND MAINTENANCE OF CONVEYOR SYSTEMS

A conveyor system is often the lifeline to a company's ability to effectively move its product in a timely fashion. The steps that a company can take to ensure that it performs at peak capacity, include regular inspections, close monitoring of motors and reducers, keeping key parts in stock, and proper training of personnel.

Increasing the service life of your conveyor system involves: choosing the right conveyor type, the right system design and paying attention to regular maintenance practices.

A conveyor system that is designed properly will last a long time with proper maintenance. Here are six of the biggest problems to watch for in overhead type conveyor systems including I-beam monorails, enclosed track conveyors and power and free conveyors.

Poor take-up adjustment: This is a simple adjustment on most systems yet it is often it is overlooked. The chain take-up device ensures that the chain is pulled tight as it leaves the drive unit. As wear occurs and the chain lengthens, the take-up extends under the force of its springs. As they extend, the spring force becomes less and the take-up has less effect. Simply compress the take-up springs and your problem goes away. Failure to do this can result in chain surging, jamming, and extreme wear on the track and chain. Take-up adjustment is also important for any conveyor using belts as a means to power rollers, or belts themselves being the mover. With poor-take up on belt-driven rollers, the belt may twist into the drive unit and cause damage, or at the least a noticeable decrease or complete loss of performance may occur. In the case of belt conveyors, a poor take-up may cause drive unit damage or may let the belt slip off of the side of the chassis.

Lack of lubrication: Chain bearings require lubrication in order to reduce friction. The chain pull that the drive experiences can double if the bearings are not lubricated. This can cause the system to overload by either its mechanical or electrical overload protection. On conveyors that go through hot ovens, lubricators can be left on constantly or set to turn on every few cycles.

Contamination: Paint, powder, acid or alkaline fluids, abrasives, glass bead, steel shot, etc. can all lead to rapid deterioration of track and chain. Ask any bearing company about the leading cause of bearing failure and they will point to contamination. Once a foreign substance lands on the raceway of a bearing or on the track, pitting of the surface will occur, and once the surface is compromised, wear will accelerate. Building shrouds around your conveyors can help prevent the ingress of contaminants. Or, pressurize the contained area using a simple fan and duct arrangement. Contamination can also apply to belts (causing slippage, or in the case of some materials premature wear), and of the motors themselves. Since the motors can generate a considerable amount of heat, keeping the surface clean is an almost-free maintenance procedure that can keep heat from getting trapped by dust and grime, which may lead to motor burnout.

Product Handling: In conveyor systems that may be suited for a wide variety of products, such as those in distribution centers, it is important that each new product be deemed acceptable for conveying before being run through the materials handling equipment. Boxes that are too small, too large, too heavy, too light, or too awkwardly shaped may not convey, or may cause many problems including jams, excess wear on conveying equipment, motor overloads, belt breakage, or other damage, and may also consume extra man-hours in terms of picking up cases that slipped between rollers, or damaged product that was not meant for materials handling.

If a product such as this manages to make it through most of the system, the sortation system will most likely be the affected, causing jams and failing to properly place items where they are assigned. It should also be noted that any and all cartons handled on any conveyor should be in good shape or spills, jams, downtime, and possible accidents and injuries may result.

Drive Train: Notwithstanding the above, involving take-up adjustment, other parts of the drive train should be kept in proper shape. Broken O-rings on a Lineshaft, pneumatic parts in disrepair, and motor reducers should also be inspected. Loss of power to even one or a few rollers on a conveyor can mean the difference between effective and timely delivery, and repetitive nuances that can continually cost downtime.

Bad Belt Tracking or Timing: In a system that uses precisely controlled belts, such as a sorter system, regular inspections should be made that all belts are traveling at the proper speeds at all times. While usually a computer controls this with Pulse Position Indicators, any belt not controlled must be monitored to ensure accuracy and reduce the likelihood of problems. Timing is also important for any equipment that is instructed to precisely meter out items, such as a merge where one box pulls from all lines at one time. If one were to be mistimed, product would collide and disrupt operation. Timing is also important wherever a conveyor must "keep track" of where a box is, or improper operation will result.

Since a conveyor system is a critical link in a company's ability to move its products in a timely fashion, any disruption of its operation can be costly. Most "downtime" can be avoided by taking steps to ensure a system operates at peak performance, including regular inspections, close monitoring of motors and reducers, keeping key parts in stock, and proper training of personnel.

Conveyor belt

A conveyor belt (or belt conveyor) consists of two or more pulleys, with a continuous loop of material - the conveyor belt - that rotates about them. One or both of the pulleys are powered, moving the belt and the material on the belt forward. The powered pulley is called the drive pulley while the unpowered pulley is called the idler. There are two main industrial classes of belt conveyors.

Those in general material handling such as those moving boxes along inside a factory and bulk material handling such as those used to transport industrial and agricultural materials, such as grain, coal, ores, etc. generally in outdoor locations. Generally companies providing general material handling type belt conveyors do not provide the conveyors for bulk material handling. In addition there are a number of commercial applications of belt conveyors such as those in grocery stores.

The belt consists of one or more layers of material they can be made out of rubber. Many belts in general material handling have two layers. An under layer of material to provide linear strength and shape called a carcass and an over layer called the cover. The carcass is often a cotton or plastic web or mesh. The cover is often various rubber or plastic compounds specified by use of the belt. Covers can be made from more exotic materials for unusual applications such as silicone for heat or gum rubber when traction is essential.

Material flowing over the belt may be weighed in transit using a beltweigher. Belts with regularly spaced partitions, known as elevator belts, are used for transporting loose materials up steep inclines. Belt Conveyors are used in self-unloading bulk freighters and in live bottom trucks. Conveyor technology is also used in conveyor transport such as moving sidewalks or escalators, as well as on many manufacturing assembly lines. Stores often have conveyor belts at the check-out counter to move shopping items. Ski areas also use conveyor belts to transport skiers up the hill.

A wide variety of related conveying machines are available, different as regards principle of operation, means and direction of conveyance, including screw conveyors, vibrating conveyors, pneumatic conveyors, the moving floor system, which uses reciprocating slats to move cargo, and roller conveyor system, which uses a series of powered rollers to convey boxes or pallets.

BELT CONVEYOR SYSTEMS

Conveyors are used as components in automated distribution and warehousing. In combination with computer controlled pallet handling equipment this allows for more efficient retail, wholesale, and manufacturing distribution. It is considered a labor saving system that allows large volumes to move rapidly through a process, allowing companies to ship or receive higher volumes with smaller storage space and with less labor expense.

Rubber conveyor belts are commonly used to convey items with irregular bottom surfaces, small items that would fall in between rollers (e.g. a sushi conveyor bar), or bags of product that would sag between rollers. Belt conveyors are generally fairly similar in construction consisting of a metal frame with rollers at either end of a flat metal bed. The belt is looped around each of the rollers and when one of the rollers is powered (by an electrical motor) the belting slides across the solid metal frame bed, moving the product. In heavy use applications the beds which the belting is pulled over are replaced with rollers.

The rollers allow weight to be conveyed as they reduce the amount of friction generated from the heavier loading on the belting. Belt conveyors can now be manufactured with curved sections which use tapered rollers and curved belting to convey products around a corner. These conveyor systems are commonly used in postal sorting offices and airport baggage handling systems. A sandwich belt conveyor uses two conveyor belts, face-to-face, to firmly contain the item being carried, making steep incline and even vertical-lift runs achievable.

Belt conveyors are the most commonly used powered conveyors because they are the most versatile and the least expensive. Product is conveyed directly on the belt so both regular and irregular shaped objects, large or small, light and heavy, can be transported successfully. These conveyors should use only the highest quality premium belting products, which reduces belt stretch and results in less maintenance for tension adjustments. Belt conveyors can be used to transport product in a straight line or through changes in elevation or direction. In certain applications they can also be used for static accumulation.

Long belt conveyors

The longest belt conveyor system in the world is in Western Sahara. It is 100 km long, from the phosphate mines of Bu Craa to the coast south of El-Aaiun. The longest conveyor system in an Airport is the Dubai International Airport baggage handling system at 92 km. It was installed by Siemens and commissioned in 2008, and has a combination of traditional belt conveyors and tray conveyors.

The longest single belt conveyor runs from Meghalaya in India to Sylhet in Bangladesh. It is about 17 km long and conveys limestone and shale at 960 tons/hr, from the quarry in India to a cement factory in Bangladesh (7 km long in India and

10 km long in Bangladesh). The conveyor was engineered by AUMUND France and Larsen & Toubro. The conveyor is actuated by 3 synchronized drive units for a total power of about 1.8 MW (2 drives at the head end in Bangladesh and 1 drive at the tail end in India).

The conveyor belt was manufactured in 300-meter lengths on the Indian side and 500-meter lengths on the Bangladesh side, and was installed onsite by NILOS India. The idlers, or rollers, of the system are unique in that they are designed to accommodate both horizontal and vertical curves along the terrain. Dedicated vehicles were designed for the maintenance of the conveyor, which is always at a minimum height of 5 meters above the ground to avoid being flooded during monsoon periods.

Screw Conveyor

A screw conveyor or auger conveyor is a mechanism that uses a rotating helical screw blade, called a "flighting", usually within a tube, to move liquid or granular materials. They are used in many bulk handling industries. Screw conveyors in modern industry are often used horizontally or at a slight incline as an efficient way to move semi-solid materials, including food waste, wood chips, aggregates, cereal grains, animal feed, boiler ash, meat and bone meal, municipal solid waste, and many others. The first type of screw conveyor was the archimedian screw, used since ancient times to pump irrigation water.

They usually consist of a trough or tube containing either a spiral blade coiled around a shaft, driven at one end and held at the other, or a "shaftless spiral", driven at one end and free at the other. The rate of volume transfer is proportional to the rotation rate of the shaft. In industrial control applications the device is often used as a variable rate feeder by varying the rotation rate of the shaft to deliver a measured rate or quantity of material into a process.

Screw conveyors can be operated with the flow of material inclined upward. When space allows, this is a very economical method of elevating and conveying. As the angle of inclination increases, the allowable capacity of a given unit rapidly decreases.

The rotating part of the conveyor is sometimes called simply an auger.

The "grain auger" is used in agriculture to move grain from trucks and grain carts into grain storage bins (from where it is later removed by gravity chutes at the bottom). A grain auger may be powered by an electric motor; a tractor, through the power take-off; or sometimes an internal combustion engine mounted on the auger. The helical flighting rotates inside a long metal tube, moving the grain upwards. On the lower end, a hopper receives grain from the truck or grain cart. A chute on the upper end guides the grain into the destination location.

The modern grain auger of today's farming communities was invented by Peter Pakosh. His grain mover employed a screw-type auger with a minimum of moving parts, a totally new application for this specific use. At Massey Harris (later Massey Ferguson), young Pakosh approached the design department in the 1940s with his auger idea, but was scolded and told that his idea was unimaginable and that once the auger aged and bent that the metal on metal would, according to a head Massey designer, "start fires all across Canada". Pakosh, however, went on to design and build a first prototype auger in 1945, and 8 years later start selling tens of thousands under the 'Versatile' name, making it the standard for modern grain augers.

A specialized form of grain auger is used to transfer grain into a seed drill, and is usually quite a lot smaller in both length and diameter than the augers used to transfer grain to or from a truck, grain cart or bin. This type of auger is known as a "drill fill". Grain augers with a small diameter, regardless of the use they are put to, are often called "pencil augers".

Other uses

Various other applications of the screw or auger conveyor include its use in snowblowers, to move snow towards an impeller, where it is thrown into the discharge chute. Combine harvesters use both enclosed and open augers to move the unthreshed crop into the threshing mechanism and to move the grain into and out of the machine's hopper. Ice resurfacers use augers to remove loose ice particles from the surface of the ice. An auger is also a central component of an injection molding machine. An auger is used in some rubbish compactors to push the rubbish into a lowered plate at one end for compaction.

Augers are also used to force food products through dies to produce pellets. These are then processed further to produce products such as bran flakes. Augers are also used in oil fields as a method of transporting rock cuttings away from the shakers to skips. Augers are also used in some types of pellet fed barbecue grills, to move fuel from a hopper into the firebox in a controlled manner. Augers are often used in machining, wherein the machine tools may include an auger to direct the swarf (scrap metal or plastic) away from the work piece.

Hoist (device)

A hoist is a device used for lifting or lowering a load by means of a drum or lift-wheel around which rope or chain wraps. It may be manually operated, electrically or pneumatically driven and may use chain, fiber or wire rope as its lifting medium. The load is attached to the hoist by means of a lifting hook.

The basic hoist has two important characteristics to define it: Lifting medium and power type. The lifting medium is either wire rope, wrapped around a drum, or load-chain, raised by a pulley with a special profile to engage the chain. The power type can be either electric motor or air motor. Both the wire rope hoist and chain hoist have been in common use since the 1800s. however; Mass production of an electric hoist did not start until the early 1900's and was first adapted by Germany.

A hoist can be built as one integral-package unit, designed for cost-effective purchasing and moderate use, or it can be built as a built-up custom unit, designed for durability and performance. The built-up hoist will be much more expensive, but will also be easier to repair and more durable. Package units are where once regarded as being designed for light to moderate usage, but since the 60's this has changed.

Built-up units are designed for heavy to severe service, but over the years that market has decreased in size since the advent of the more durable packaged hoist. A machine shop or fabricating shop will use an integral-package hoist, while a Steel Mill or NASA would use a built-up unit to meet durability, performance, and repairability requirements. NASA has also seen a change in the use of package hoists. The NASA Astronaut training pool for example utilizes cranes with packaged hoists.

WIRE ROPE HOIST OR CHAIN HOIST

More commonly used hoist in today's worldwide market is an electrically powered hoist. These are either the chain type or the wire rope type. Demag Cranes & Components Corp. was one of the first companies in the world to mass-produce hoists. The first units large in size date back in 1819 and was powered by steam.

Now many hoists are package hoists, built as one unit in a single housing, generally designed for ten-year life, but the life calculation is based on an industry standard when calculating actual life. See the Hoists Manufacturers Institute site for true life calculation witch is based on load and hours used. In today's modern world for the North American market there are a few governing bodies for the industry.

The Overhead Alliance is a group that represents Crane Manufacturers Association of America (CMAA), Hoist Manufacturers Institute (HMI), and Monorail Manufacturers Association (MMA). These product counsels of the Material Handling Industry of America have joined forces to create promotional materials to raise the awareness of the benefits to overhead lifting. The members of this group are marketing representatives of the member companies.

Common small portable hoists are of two main types, the chain hoist or chain block and the wire rope or cable type. Chain hoists may have a lever to actuate the hoist or have a loop of operating chain that one pulls through the block (known traditionally as a chain fall) which then activates the block to take up the main lifting chain.

A hand powered hoist with a ratchet wheel is known as a "ratchet lever hoist" or, colloquially, a "Come-A-Long". The original hoist of this type was developed by Abraham Maasdam of Deep Creek, Colorado about 1919, and later commercialized by his son, Felber Maasdam, about 1946. It has been copied by many manufacturers in recent decades. A similar heavy duty unit with a combination chain and cable became available in 1935 that was used by railroads, but lacked the success of the cable only type units.

Ratchet lever hoists have the advantage that they can usually be operated in any orientation, for pulling, lifting or binding. Chain block type hoists are usually suitable only for vertical lifting.

For a given rated load wire rope is lighter in weight per unit length but overall length is limited by the drum diameter that the cable must be wound onto. The lift chain of a chain hoist is far larger than the liftwheel over which chain may function. Therefore, a high-performance chain hoist may be of significantly smaller physical size than a wire rope hoist rated at the same working load.

Both systems fail over time through fatigue fractures if operated repeatedly at loads more than a small percentage of their tensile breaking strength. Hoists are often designed with internal clutches to limit operating loads below this threshold. Within such limits wire rope rusts from the inside outward while chain links are markedly reduced in cross section through wear on the inner surfaces. Regular lubrication of both tensile systems is recommended to reduce frequency of replacement. High speed lifting, greater than about 60 feet per minute (18.3 m/min), requires wire rope wound on a drum, because chain over a pocket wheel generates fatigue-inducing resonance for long lifts.

The unloaded wire rope of small hand powered hoists often exhibits a snarled "set", making the use of a chain hoist in this application less frustrating, but heavier. In addition, if the wire in a wire hoist fails, it can whip and cause injury, while a chain will simply break.

Construction hoists

Also known as a Man-Lift, Buckhoist, temporary elevator, Alimak or construction elevator, this type of hoist is commonly used on large scale construction projects, such as high-rise buildings or major hospitals. There are many other uses for the construction elevator. Many other industries use the buckhoist for full time operations. The purpose being to carry personnel, materials, and equipment quickly between the ground and higher floors, or between floors in the middle of a structure.

The construction hoist is made up of either one or two cars (cages) which travel vertically along stacked mast tower sections. The mast sections are attached to the structure or building every 25 feet (7.62 m) for added stability. For precisely controlled travel along the mast sections, modern construction hoists use a motorized rack-and-pinion system that climbs the mast sections at various speeds.

While hoists have been predominantly produced in Europe and the United States, China is emerging as a manufacturer of hoists to be used in Asia.

In the United States and abroad, General Contractors and various other industrial markets rent or lease hoists for a specific projects. Rental or leasing companies provide erection, dismantling, and repair services to their hoists to provide General Contractors with turnkey services. Also the rental and leasing companies can provide parts and service for the elevators that are under contract.

Mine hoists

In underground mining a hoist or winder is used to raise and lower conveyances within the mine shaft. Human, animal and water power were used to power the mine hoists documented in Agricola's De Re Metallica, published in 1556. Stationary steam engines were commonly used to power mine hoists through the 19th century and into the 20th, as at the Quincy Mine, where a 4-cylinder cross-compound corliss engine was used. Modern hoists are powered using electric motors, historically with direct current drives utilizing solid-state converters (thyristors), however modern large hoists use alternating current drives that are variable frequency controlled. There are three principal types of hoists used in mining applications, Drum Hoists, Friction (or Kope) hoists and Blair multi-rope hoists.

10

MECHANICAL ENGINEERING DESIGN

Mechanical engineering is an engineering branch that combines engineering physics and mathematics principles with materials science to design, analyze, manufacture, and maintain mechanical systems. It is one of the oldest and broadest of the engineering branches.

The mechanical engineering field requires an understanding of core areas including mechanics, dynamics, thermodynamics, materials science, structural analysis, and electricity. In addition to these core principles, mechanical engineers use tools such as computer-aided design (CAD), computer-aided manufacturing (CAM), and product lifecycle management to design and analyze manufacturing plants, industrial equipment and machinery, heating and cooling systems, transport systems, aircraft, watercraft, robotics, medical devices, weapons, and others. It is the branch of engineering that involves the design, production, and operation of machinery.

Mechanical engineering emerged as a field during the Industrial Revolution in Europe in the 18th century; however, its development can be traced back several thousand years around the world. In the 19th century, developments in physics led to the development of mechanical engineering science. The field has continually evolved to incorporate advancements; today mechanical engineers are pursuing developments in such areas as composites, mechatronics, and nanotechnology. It also overlaps with aerospace engineering, metallurgical engineering, civil engineering, electrical engineering, manufacturing engineering, chemical engineering, industrial engineering, and other engineering disciplines to varying amounts. Mechanical engineers may also work in the field of biomedical engineering, specifically with biomechanics, transport phenomena, biomechatronics, bionanotechnology, and modelling of biological systems.

DUTIES OF MECHANICAL ENGINEERS

Mechanical engineers research, design, develop, build, and test mechanical and thermal devices, including tools, engines, and machines.

Mechanical engineers typically do the following:

Analyze problems to see how mechanical and thermal devices might help solve the problem.

Design or redesign mechanical and thermal devices using analysis and computer-aided design.

Develop and test prototypes of devices they design.

Analyze the test results and change the design as needed.

Oversee the manufacturing process for the device.

Mechanical engineers design and oversee the manufacturing of many products ranging from medical devices to new batteries. They also design power-producing machines such as electric generators, internal combustion engines, and steam and gas turbines as well as power-using machines, such as refrigeration and air-conditioning systems.

Like other engineers, mechanical engineers use computers to help create and analyze designs, run simulations and test how a machine is likely to work.

SUBDISCIPLINES OF MECHANICAL ENGINEERING

The field of mechanical engineering can be thought of as a collection of many mechanical engineering science disciplines. Several of these subdisciplines which are typically taught at the undergraduate level are listed below, with a brief explanation and the most common application of each.

Some of these subdisciplines are unique to mechanical engineering, while others are a combination of mechanical engineering and one or more other disciplines. Most work that a mechanical engineer does uses skills and techniques from several of these subdisciplines, as well as specialized subdisciplines.

Mechanics

Mechanics is, in the most general sense, the study of forces and their effect upon matter. Typically, engineering mechanics is used to analyze and predict the acceleration and deformation (both elastic and plastic) of objects under known forces (also called loads) or stresses. Subdisciplines of mechanics include

Statics, the study of non-moving bodies under known loads, how forces affect static bodies.

Dynamics the study of how forces affect moving bodies. Dynamics includes kinematics (about movement, velocity, and acceleration) and kinetics (about forces and resulting accelerations).

Mechanics of materials, the study of how different materials deform under various types of stress

Fluid mechanics, the study of how fluids react to forces

Kinematics, the study of the motion of bodies (objects) and systems (groups of objects), while ignoring the forces that cause the motion. Kinematics is often used in the design and analysis of mechanisms.

Continuum mechanics, a method of applying mechanics that assumes that objects are continuous (rather than discrete)

Mechanical engineers typically use mechanics in the design or analysis phases of engineering. If the engineering project were the design of a vehicle, statics might be employed to design the frame of the vehicle, in order to evaluate where the stresses will be most intense. Dynamics might be used when designing the car's engine, to evaluate the forces in the pistons and cams as the engine cycles. Mechanics of materials might be used to choose appropriate materials for the frame and engine. Fluid mechanics might be used to design a ventilation system for the vehicle (see HVAC), or to design the intake system for the engine.

WHAT DO MECHANICAL ENGINEERS DO?

Mechanical engineering combines creativity, knowledge and analytical tools to complete the difficult task of shaping an idea into reality.

This transformation happens at the personal scale, affecting human lives on a level we can reach out and touch like robotic prostheses. It happens on the local scale, affecting people in community-level spaces, like with agile interconnected microgrids. And it happens on bigger scales, like with advanced power systems, through engineering that operates nationwide or across the globe. Mechanical engineers have an enormous range of opportunity and their education mirrors this breadth of subjects. Students concentrate on one area while strengthening analytical and problem-solving skills applicable to any engineering situation.

Disciplines within mechanical engineering include but are not limited to:

- Acoustics
- Aerospace

- Automation
- Automotive
- Autonomous Systems
- Biotechnology
- Composites
- Computer Aided Design (CAD)
- Control Systems
- Cyber security
- Design
- Energy
- Ergonomics
- Human health
- Manufacturing and additive manufacturing
- Mechanics
- Nanotechnology
- Production planning
- Robotics
- Structural analysis

Technology itself has also shaped how mechanical engineers work and the suite of tools has grown quite powerful in recent decades. Computer-aided engineering (CAE) is an umbrella term that covers everything from typical CAD techniques to computer-aided manufacturing to computer-aided engineering, involving finite element analysis (FEA) and computational fluid dynamics (CFD). These tools and others have further broadened the horizons of mechanical engineering.

WHAT CAREERS ARE THERE IN MECHANICAL ENGINEERING?

Society depends on mechanical engineering. The need for this expertise is great in so many fields, and as such, there is no real limit for the freshly minted mechanical engineer. Jobs are always in demand, particularly in the automotive, aerospace, electronics, biotechnology, and energy industries.

Here are a handful of mechanical engineering fields.

In statics, research focuses on how forces are transmitted to and throughout a structure. Once a system is in motion, mechanical engineers look at dynamics, or what velocities, accelerations and resulting forces come into play. Kinematics then examines how a mechanism behaves as it moves through its range of motion. Materials science delves into determining the best materials for different applications. A part of that is materials strength—testing support loads, stiffness, brittleness and other properties—which is essential for many construction, automobile, and medical materials.

How energy gets converted into useful power is the heart of thermodynamics, as well as determining what energy is lost in the process. One specific kind of energy, heat transfer, is crucial in many applications and requires gathering and analyzing temperature data and distributions. Fluid mechanics, which also has a variety of applications, looks at many properties including pressure drops from fluid flow and aerodynamic drag forces.

Manufacturing is an important step in mechanical engineering. Within the field, researchers investigate the best processes to make manufacturing more efficient. Laboratory methods focus on improving how to measure both thermal and mechanical engineering products and processes. Likewise, machine design develops equipment-scale processes while electrical engineering focuses on circuitry. All this equipment produces vibrations, another field of mechanical engineering, in which researchers study how to predict and control vibrations.

Engineering economics makes mechanical designs relevant and usable in the real world by estimating manufacturing and life cycle costs of materials, designs, and other engineered products. The essence of engineering is problem solving. With this at its core, mechanical engineering also requires applied creativity—a hands on understanding of the work involved—along with strong interpersonal skills like networking, leadership, and conflict management. Creating a product is only part of the equation; knowing how to work with people, ideas, data, and economics fully makes a mechanical engineer.

FUTURE OF MECHANICAL ENGINEERING

Breakthroughs in materials and analytical tools have opened new frontiers for mechanical engineers. Nanotechnology, biotechnology, composites, computational fluid dynamics (CFD), and acoustical engineering have all expanded the mechanical engineering toolbox. Nanotechnology allows for the engineering of materials on the smallest of scales. With the ability to design and manufacture down to the elemental level, the possibilities for objects grows immensely. Composites are another area where the manipulation of materials allows for new manufacturing opportunities. By combining materials with different characteristics in innovative ways, the best of each material can be employed and new solutions found. CFD gives mechanical engineers the opportunity to study complex fluid flows analyzed with algorithms. This allows for the modeling of situations that would previously have been impossible. Acoustical engineering examines vibration and sound, providing the opportunity to reduce noise in devices and increase efficiency in everything from biotechnology to architecture.

INTRODUCTION TO MECHANICAL ENGINEERING DESIGN

Mechanical engineering design includes all mechanical design, but it is a broader study because it includes all the disciplines of mechanical engineering, such as the thermal fluids and heat transfer sciences too.

Aside from the fundamental sciences which are required, the first studies in mechanical engineering design are in mechanical design, and that is the approach taken in this course.

Steps of the Design Process

Recognize the Need

The first step is to establish the ultimate purpose of the project. Often, this is in the form of a general statement of the client's dissatisfaction with a current situation.

Example – "There is too much damage to bumpers in low-speed collisions."

This is a general statement that does not comment on the design approach to the problem. It does not say that the bumper should be stronger or more flexible.

Recognition and phrasing of the need are often very creative acts because the need may only be a sensing that something is not right. For this reason, sensitive people are generally more creative.

Example – the need to do something about a food packaging machine may be indicated by the noise level, variation in package weights, or by slight but perceptible variations in the quality of the packaging.

Problem Definition

This is one of the most critical steps of the design process. There is an iteration between the definition of the problem and the recognition of need. Often the true problem is not what it first seems. The problem definition is more specific than recognizing the need. For instance, if the need is for cleaner air, the problem might be that of reducing the dust discharge from power-plant stacks, or reducing the quantity of irritants from automotive exhausts, or means for quickly extinguishing forest fires. The problem definition must include all the specifications for the thing that is to be designed. Anything which limits the designer's freedom of choice is a specification. It is imperative to write a formal problem statement which expresses what the design is to accomplish include: objectives and goals (musts, must nots; wants, don't wants) constraints criteria used to evaluate the design.

PHASES AND INTERACTIONS OF THE DESIGN PROCESS

The complete design process, from start to finish, is often outlined as in Fig. The process begins with an identification of a need and a decision to do something about it. After many iterations, the process ends with the presentation of the plans for satisfying the need. Depending on the nature of the design task, several design phases may be repeated throughout the life of the product, from inception to termination.

For example, the need to do something about a food-packaging machine may be indicated by the noise level, by a variation in package weight, and by slight but perceptible variations in the quality of the packaging or wrap.

There is a distinct difference between the statement of the need and the definition of the problem. The definition of problem is more specific and must include all the specifications for the object that is to be designed. The specifications are the input and output quantities, the characteristics and dimensions of the space the object must occupy, and all the limitations on these quantities. We can regard the object to be designed as something in a black box. In this case we must specify the inputs and outputs of the box, together with their characteristics and limitations. The specifications define the cost, the number to be manufactured, the expected life, the range, the operating temperature, and the reliability. Specified characteristics can include the speeds, feeds, temperature limitations, maximum range, expected variations in the variables, dimensional and weight limitations, etc.

There are many implied specifications that result either from the designer's particular environment or from the nature of the problem itself. The manufacturing processes that are available, together with the facilities of a certain plant, constitute restrictions on a designer's freedom, and hence are a part of the implied specifications. It may be that a small plant, for instance, does not own cold-working machinery.

Knowing this, the designer might select other metal-processing methods that can be performed in the plant. The labor skills available and the competitive situation also constitute implied constraints. Anything that limits the designer's freedom of choice is a constraint. Many materials and sizes are listed in supplier's catalogs, for instance, but these are not all easily available and shortages frequently occur. Furthermore, inventory economics requires that a manufacturer stock a minimum number of materials and sizes. This example is for a case study of a power transmission that is presented throughout this text.

The synthesis of a scheme connecting possible system elements is sometimes called the invention of the concept or concept design. This is the first and most important step in the synthesis task. Various schemes must be proposed, investigated, and quantified in terms of established metrics. As the fleshing out of the scheme progresses, analyses must be performed to assess whether the system performance is satisfactory or better, and, if satisfactory, just how well it will perform. System schemes that do not survive analysis are revised, improved, or discarded. Those with potential are optimized to determine the best performance of which the scheme is capable. Competing schemes are compared so that the path leading to the most competitive product can be chosen. Figure shows that synthesis and analysis and optimization are intimately and iteratively related.

We have noted, and we emphasize, that design is an iterative process in which we proceed through several steps, evaluate the results, and then return to an earlier phase of the procedure. Thus, we may synthesize several components of a system, analyze and optimize them, and return to synthesis to see what effect this has on the remaining parts of the system. For example, the design of a system to transmit power requires attention to the design and selection of individual components (e.g., gears, bearings, shaft).

However, as is often the case in design, these components are not independent. In order to design the shaft for stress and deflection, it is necessary to know the applied forces. If the forces are transmitted through gears, it is necessary to know the gear specifications in order to determine the forces that will be transmitted to the shaft. But stock gears come with certain bore sizes, requiring knowledge of the necessary shaft diameter.

Clearly, rough estimates will need to be made in order to proceed through the process, refining and iterating until a final design is obtained that is satisfactory for each individual component as well as for the overall design specifications. Throughout the text we will elaborate on this process for the case study of a power transmission design. Both analysis and optimization require that we construct or devise abstract models of the system that will admit some form of mathematical analysis. We call these models mathematical models. In creating them it is our hope that we can find one that will simulate the real physical system very well. As indicated in Fig. evaluation is a significant phase of the total design process. Evaluation is the final proof of a successful design and usually involves the testing of a prototype in the laboratory. Here we wish to discover if the design really satisfies the needs. Is it reliable? Will it compete successfully with similar products? Is it economical to manufacture and to use? Is it easily maintained and adjusted? Can a profit be made from its sale or use? How likely is it to result in product-liability lawsuits? And is insurance easily and cheaply obtained? Is it likely that recalls will be needed to replace defective parts or systems?

Communicating the design to others is the final, vital presentation step in the design process. Undoubtedly, many great designs, inventions, and creative works have been lost to posterity simply because the originators were unable or unwilling to explain their accomplishments to others. Presentation is a selling job. The engineer, when presenting a new solution to administrative, management, or supervisory persons, is attempting to sell or to prove to them that this solution is a better one. Unless this can be done successfully, the time and effort spent on obtaining the solution have been largely wasted. When designers sell a new idea, they also sell themselves. If they are repeatedly successful in selling ideas, designs, and new solutions to management, they begin to receive salary increases and promotions; in fact, this is how anyone succeeds in his or her profession.

INTERACTION DESIGN ACTIVITIES

As the previous activity shows, in order for an interactive product to do a good job it must be designed with the user in mind. Indeed, user-centred design is a core approach of interaction design, meaning that every good interactive product is designed around the users, their environment and their activities, so that it is fit for purpose.

In other words, you conducted the four fundamental activities that make up the interaction design process – establishing requirements, designing alternatives, prototyping designs, and evaluating prototypes.

Establishing requirements – a requirement is a need that a particular interactive product must be able to satisfy. Establishing what is required of the product is essential to ensure that the interaction is the best possible fit for the user, both in terms of what the user needs to do with the product and how they experience the interaction. Requirements will depend on the characteristics of the user, the activities the user will perform using the product, and the environment in which the user interacts with the product. In the example of a phone or remote control, requirements are shaped by the need to use the device (e.g. mobile phone) to do certain activities (to make phone calls), given the size and mobility of the user's hands (bigger than standard or fingerless) and the user's physical environment (ski slope).

Designing alternatives – coming up with alternative designs enables designers to explore different ways of interpreting and satisfying the requirements for a particular interactive product. This is an essential and highly creative part of the process. In the phone and remote control example, this activity began when you started jotting down alternatives for the controls. Design ideas should be informed by fundamental design principles that derive from what we know about how our minds and bodies work.

Prototyping designs – once interaction designers have identified a number of possible ideas, they need to figure out which ones have the potential to work best for the users, their activities and their environment. To do this, designers need to prototype the most promising design ideas to make a first, often rough, model so that they can try them out. In the example of the phone or remote control, as you thought of different designs, you were also prototyping them by drawing the alternative interfaces you thought of. Prototyping can also be used to explore different aspects of a design.

Evaluating prototypes – evaluation enables designers to assess the limitations of a particular design, to find out to what extent a prototype meets requirements that have already been identified, to identify requirements that have not already emerged, and to establish what changes need to be made so that requirements are met.

In your interaction design exercise, you performed a rough evaluation of your paper-based prototypes of a phone or remote interface by trying to 'interact' with your designs while wearing the gloves or socks. Therefore, while this was a task that was relatively easy to execute, it had all the elements of what we consider to be the fundamental activities in interaction design. In this course you will be introduced to different ways of achieving better designs that are informed by the needs of users.

DESIGN TOOLS AND RESOURCES

Being an engineer is as demanding as it is exciting and rewarding. A widening skill gap for engineers of all types has ensured there are more projects than qualified people to build them. If your department is overstaffed, if you are overworked and expected to know every engineering discipline within your development team, you are not alone. But there are tools available to help streamline your processes so you can create more, and fight with circuit design less. Here are the best tools for CAD, CAM, simulation, PCB layout, and coding to make your busy, overworked life just a little easier. As computer programs and mobile apps take over our lives, one of the great things about being a mechanical engineer is that there are still aspects of the job that allow — and sometimes require — us to do things by hand, the "old fashioned" way.

Engineering Design Services

Because modern technology is so complex, it is now near about impossible for an individual to handle design and development of a new product single handedly. It takes a team of designers and engineers to successfully manufacture a new product. In order to achieve success, the design process must be planned carefully and executed systematically. Specifically, the engineering design process must integrate the many different aspects of designing in such a way that the whole process becomes logical and comprehensible. In addition, most of the CAD / CAE software available today is complex. It takes special training to understand and utilize the harness the real power of these applications. Secondly, the cost of acquiring these tools is rather expensive. Many companies that need to develop products therefore outsource their engineering design requirements to agencies that provide such services. There are also times when the company has trained manpower and good CAD / CAE infrastructure, but the people working there are simply overloaded with work. In such cases too, companies avail engineering design services to share their work load when they are hard pressed for time.

In developing nations (like India), engineering design services are the need of the hour. There are many small and medium scale industries that do not have the necessary infrastructure to thoroughly evaluate their product ideas and reach the manufacturing stage, and reputed design services companies are in good demand to assist them.

Granted, drafting boards and T-squares have been replaced by CAD programs, complex modeling and analysis is done in a matter a minutes with FEA software, and plug-and-play configuration programs have greatly reduced the number of iterations it takes to size and select a ball screw or linear bearing. But you can't disassemble a gearbox without a bearing puller (at least, not easily and efficiently), and you can't measure the ID of a radial bearing without a set of calipers. So, for those times when you get to step away from the computer and phone and get back to hands-on engineering and design, here are seven tools that will make the job easier.

Basic Toolbox

Most engineers don't have the space or budget to keep a full-fledged toolbox at their desk. But even if you can't have all your favorite tools at your fingertips, there's no excuse for not having a basic set of screwdrivers, hex keys, wrenches, needle nose pliers, and a few other job-specific tools at your desk.

Scientific Calculator

Although every phone, tablet, and computer comes with a built-in calculator these days, a good scientific calculator (with RPN logic if you're old-school) cannot be replaced. There's some disagreement in the engineering community about whether a graphing calculator is necessary these days, since Excel can do pretty much anything a graphing calculator can do. But if you're working with complex calculations, you can't beat the additional functionality and expanded command stack that a graphing calculator offers.

Engineering pad/graph Paper

It's one of the most basic tools of engineering, but nothing beats it for sketching designs or documenting calculations.

Design Tools

Working as a mechanical engineer in the United States is exciting, rewarding but also demanding. An increasing skill gap has ensured that there are more projects out there than qualified professionals to complete them.

As your role becomes more demanding and you are expected to have a lot of knowledge, sometimes working in an interdisciplinary role, this can seem overwhelming. Fortunately, there are a multitude of design tools available today to help streamline your workflow so you can design more, and be a more efficient engineer in general. Below are a few of the best tools for CAD, CAM, PCB layout, simulation and programming to hopefully alleviate some of the pressure you are experiencing as you are being overworked in your job!

Computer Aided Design (CAD)

Solid Works

Solid works the most popular CAD/CAE software in the world whether you work with 2-D or 3-D designs. This software is very versatile, it supports a multitude of industries, including industrial and mechanical engineering in aerospace, architecture, augmented reality, automotive, energy, medical, shipbuilding, technology, and many more.

Whether you work with 2D or 3D designs, Solidworks has been and continues to be the leading CAD/CAE software on the market, and for good reason. The software is versatile, supporting industrial design and mechanical engineering in automotive, aerospace, shipbuilding, architecture, medical, energy, technology, augmented reality, and more.

Each year, the product gets better. In the new release, Solidworks offers a range of new technology to streamline large and complex workflow applications, user interface design, and collaboration. It also offers a development environment if you are working with contractors (free with the design for manufacturing solution), which enables collaboration across design, communication, validation, and manufacturing. With this, since Solidworks is such a broadly used tool, it offers fantastic plug-ins and compatibility with a variety of milling machines and 3D printers. The software supports simulation, automated report generation, error detection, SolidCAM (a comprehensive CAM plugin tool) and more, with tailored features based on industry.

This software gets updated every year, and in its new releases SolidWorks continues to offer the newest technology in order to streamline complex and large workflow applications, improve collaboration and the design of user interfaces. If you happen to be working with contractors often, it offers a development environment which comes at no extra cost with the "design for manufacturing" solution, and this allows you to collaborate on design, communication, manufacturing and validation. Using this array of features, SolidWorks has became a widely used program, it is compatible with a wide range of 3-D printers and milling machines while also offering some fantastic plug-ins. It also supports generating automated reports, detecting errors, simulation, SolidCAM (a CAM plugin) and more, and it can be tailored with features depending on the industry that you work in.

Catia

Solidworks is a phenomenal CAD tool, but it does have a learning curve. If you're an entrepreneur or startup, Dassault Systemes (the manufacturer of Solidworks, among other engineering software tools) offers free training to help you learn the tool.

That being said, and alternative version of SolidWorks that is a bit more affordable and offers the same platform for architecture, automotive, industrial equipment or shipbuilding can be found in CATIA. CATIA stands for (Computer Aided Three-dimensional Interactive Application), this is made by the same company as SolidWorks. CATIA is a CAD/CAM/CAE program that is still a useful software tool while being cheaper than SolidWorks. If you are not a fan of the Windows platform, you can check out libreCAD as an option for a useful open source, 2-D software.

That said, if you're looking for a more affordable platform for automotive, shipbuilding, architecture, or industrial equipment, look into CATIA (Computer-Aided Three-Dimensional Interactive Application), also by Dassault Systemes. CATIA is a CAD, CAM, and CAE software tool that is more affordable than Solidworks, and still a solid program if you're within its intended industry. There is no "best" CAM software per se. The best tool for you depends largely on the application. Your CAD software is also a huge determining factor, and is another reason using Solidworks or AutoCAD is a good idea – because the tools offer compatibility with a wide range of programs.

That said, here are a few of the best available now:

If you purchase Solidworks, SolidCAM is a plugin that integrates directly with your CAD designs. It is also compatible with Inventor. SolidCAM offers exceptional features, such as automated fine-tuning of your design, automatic updates to your tool paths if something in the CAD design changes, and automated geometry definitions.

Next on the list is CATIA. This tool was mentioned above for strong CAD functionality, but it is also one of the best CAM software tools available. It's an all-in-one CAD, CAM, CAE tool, enabling you to create, validate, and manufacture your design using a single tool. Last but not least, Autodesk HSM Works is a popular plug-in for Windows; and Autodesk Fusion 360 is another leading standalone tool, also available for Mac.

CAM Software and 3D Printing

All of this conversation begs the question if CAM software is obsolete. 3D printing has taken the world by storm. Where it was once necessary to use CNC milling machines for manufacturing, most prototyping can be done with additive manufacturing, or 3D printing, today. The choice as to which best suits depends on the product you're making and your personal preference.

If you are thinking of saying so long to the milling machine in exchange for the 3D printer, you will enjoy newfound simplicity in prototyping and fabrication. Most, if not all, mainstream software offers 3D printer compatibility. If you use a program like Solidworks to create your design, simply export it to a quality 3D printer, relax, and enjoy your new prototype.

While 3D printing can absolutely be a go-to for rapid prototyping, it has shortcomings. 3D printers aren't easy to use, particularly if your design is intricate. 3D-printed objects are usually fine prototypes, but not sturdy enough to be sold as a final product unless you sell in that niche. Furthermore, 3D printers can't do everything. In fact, some engineers 3D-print objects, then further detail the design using a milling machine. All-in-all, milling machines are not yet obsolete, but it's probably helpful to know how to use it in conjunction with a 3D printer.

Construction Software

Using a top-of-the-line construction software can take your workflow to the next level. This type of program will allow you to make changes to your plans and collaborate with the other people involved with the project in real time, and this can prove to be a huge advantage. This can help to mitigate any unexpected delays or disastrous mistakes that could hinder the overall progress of the job or project. If you can greatly speed up your workflow, simplify communication with other parties/vendors and enable easy collaboration between everyone, this will vastly improve the timeline of the project. An example of this invaluable software is BIM (Business Information Modelling), which is available in Autodesk Revit, which allows the generation and exchange of data and information between the various project parties. It is an indispensable part of the construction decision-making process and without a doubt one of the essential tools for a mechanical engineer.

3-D Printers

When 3-D printers came out as tools for a mechanical engineer, and burst onto the scene, it was somewhat of a game changer. 3-D printers simplified a lot of tasks performed by mechanical engineers and they were a huge convenience to them. The fact that a mechanical engineer can easily create a 3-D model of the object that they are designing is an invaluable tool to have at their disposal.

3-D printers are at the front line of cutting edge technology available to mechanical engineers at the moment, even working mechanical engineers 5 to 10 years ago did not have access to them. As is the case with all new technology it was quite expensive at the start, now it has reached a critical point where although the technology is still futuristic, it has become affordable and a top of the range printer will only set you back a few hundred dollars.

Mechanical engineers use these 3-D printers to bring their ideas to life, so to speak, and more importantly showcase them to potential clients and investors. In minutes, a scaled-down model of your current design can be sitting in front of you. Not only is this an incredible amount of fun, but it allows greater creative freedom and a greater understanding of the final product throughout the design process. Previously, in the dark days where 3-D printers were not an option, mechanical engineers struggled trying to convey their ideas and concepts to clients and investors. It was hard to show the client the design of the product on a screen, where it is not a tangible product that you can feel or touch. Having comprehensive knowledge of how to use and operate a 3-D printer will be essential to your success as a mechanical engineer.

Internet of Things

The Internet of Things is a common platform used to integrate a variety of objects and pieces of equipment within the "IoT" itself. This is a critical skill for a mechanical engineer to have, it will allow you to streamline manufacturing and operations processes while understanding them a little better, and collect vast amounts of data quicker than ever before. This aspect of mechanical engineering and its future is bright and expanding very quickly. The final outcome or result of this field is impossible to predict and has endless opportunities associated with it, which solidifies its place on this list of the best tools for a mechanical engineer.

Electrical Design and Simulation

Webench

WEBENCH is known as a favorite in the electrical design and simulation niche, and also a classic. It is made by Texas Instruments, the tool is easily set up, intuitive and powerful. You can design a wide range of electrical applications, like developing signal chain and clock layout and power designs.

Power Designer

Also made by WEBENCH, Power Designer stakes its claim to be the most powerful end-end design tool in the field. That aside, it is a very solid tool to use. It allows the user to create a multitude of circuits and simulate them straight away. When you're done, WEBENCH will automatically generate EBOM and schematic while allowing you to choose different parts when you hover over the design. Another benefit is that it is a user friendly program for mechanical engineers with any amount of experience, while also being affordable.

PCB Layout

Autodesk Eagle

Autodesk Eagle (formerly known as EagleCAD) is the best PCB layout tool in the industry. This became the most popular PCB platform in the field, and was then bought out by Autodesk a couple of years ago and has continued its brilliance. Eagle is a program that does not only allow you to complete PCB layouts, you can also insert your designs into your 3-D models. You can also take out bits of unused board, route your wires and rotate your boards quite easily.

Altium

Altium is one of the main competitors for Eagle, and is another widely used program for its affordable price and robust features. If you are in a senior position or a manager of an electronic lab, Eagle will have all of the usability and power you need at an affordable price.

Coding

Wyliodrin

There is no way to talk about designing platforms without having to confront code, and coding is definitely one of the best tools a mechanical engineer can have in his bag.

Teaching yourself to code is like learning a new language, Coursera, edX and Udemy are all free online coaching courses. These offer free and paid options, with the paid options providing a certificate at completion. Unfortunately an in-depth knowledge of software design is essential in the modern world.

An alternative to coding can be found in Visual Programming Tools, this allows users to program devices using a drag-and-drop interface which is a lot easier to understand. They don't have to deal with variables or any of the extensive work that it takes to build commands and learn the syntax of a new platform.

These Visual Programming Tools are getting more popular and as a result of this mini programs have became available. Similar to CAM programs, the best program depends on its application, Raspberry Pi and Arduino are two examples of these programs. There are a multitude of options in the free open source category if you want to try them out before you purchase.

Wyliodrin is one of the most interesting programs out there today. It is a comprehensive tool that allows users to create programs in a variety of different coding languages.

The user interface is similar to playing a game of Tetris, and it displays the code source as you progress. It has a wide scope of compatibility; Raspberry Pi, Arduino, Grove and Beaglebone and several offerings from Intel are some examples of compatible hardware.

THE DESIGN ENGINEERS' PROFESSIONAL RESPONSIBILITIES

A design engineer creates and implements product processes and designs using computer-aided design (CAD) software. Design engineers research opportunities for new products and create prototypes. They design manufacturing processes and products with various criteria in mind, including cost-effectiveness, user experience and environmental standards. The role of a design engineer is also to analyze data from tests on prototypes and craft progress reports.

Design engineer responsibilities include:

Designing manufacturing processes and products

Assessing product usability and safety

Writing reports and presenting results to managers and customers

Design engineers study, research and develop ideas for new products and the systems used to make them. They also modify existing products or processes to increase efficiency or improve performance. They work on almost every consumer product imaginable for large-scale production, from telephones and medical equipment to kitchen appliances and car engines.

Design engineers are not only concerned with making products that look good and are easy and safe to use: they are also concerned with ensuring that the product can be made cost-effectively and efficiently. Typical tasks include:

Studying a design brief.

Thinking of possible design solutions.

Researching whether the design will work and be cost-effective.

Assessing the usability, environmental impact and safety of a design.

Using computer-aided design (CAD) and computer-assisted engineering (CAE) software to create prototypes.

Collecting and analysing data from tests on prototypes.

Modifying designs and retesting them.

Writing regular progress reports and presenting them to project managers and clients.

With experience, your career can lead to senior design positions in larger organisations and then on to posts as creative director. Design engineers can also move into management roles such as project management and new business development.

Requirements

Previous experience as a Design Engineer or a similar role

Proven experience using CAD software; knowledge of SolidWorks is a plus

Good understanding of safety standards and environmental impact of a design

Ability to present in front of stakeholders and managers

Great computational and spatial ability

Excellent oral and written communication

Attention to detail

Degree in Mechanical Engineering or a similar field

Typical Employers of Design Engineers

Design consultancy firms

Large manufacturing companies

Biomedical companies

Engineering companies

Consumer goods manufacturers

Standards and Codes

Standards are an important part of our society, serving as rules to measure or judge capacity, quantity, content, extent, value and quality. Some standards take the form of an actual item such as the atomic clock which serves as the reference for measuring time throughout the world. Others set criteria for use and practice in industry and for products used in everyday life. This introduction, however, deals primarily with standards that set a level of adequacy for structures and machines. It is these standards, above all others, which must be addressed before any engineering design project can be started.

A technical standard is an established norm or requirement. It is usually a formal document that establishes uniform engineering or technical criteria, methods, processes and practices. Standards allow for interchangeability of parts, system interoperability, and they ensure quality, reliability and safety.

How are Standards Developed?

The International Standards Organization (ISO) coordinates standards world-wide. Under this umbrella are representative organizations from many countries; for the U.S., it is the American National Standards Institute(for a listing of some organizations and their acronyms see Table 1). ANSI coordinates many national technical standards organizations. Within ANSI guidelines, each standards-writing and -issuing organization has its own method for developing standards. The American Society of Mechanical Engineers is one such approved organization, and what follows is a brief description of their standards-development process. After a new standard is suggested, the Council on Codes and Standards decides if ASME should investigate it. If the Council decides the project is worthy, it then determines the scope of the project and assigns it to the appropriate committee. The committee then develops the criteria needed to address the scope and purpose of a code or standard. Finally, after many drafts and votes, the standard is accepted by the committee.

From there it is announced in ASME's Mechanical Engineering Magazine and the ANSI Reporter so that the public can review and comment on the standard. If a comment can not be resolved by the proposing committee, then there "shall be a system of hearings and appeals" to settle the matter. How does the numbering system for standards work? There are almost as many different ways of numbering standards as there are standards-issuing organizations, although many follow a similar format. First comes the acronym of the organization which issued the standard. For example, ASTM comes before those standards originating from the American Society for Testing and Materials. This is usually followed by a letter designation that denotes the general classification of the standard. ASTM uses letters to denote certain materials:

A — ferrous metals and products

B — nonferrous metals and products

C — cementitious, ceramic, concrete and masonry materials

D — miscellaneous materials and products

E — miscellaneous subjects

F — end-use materials and products

AG — corrosion, deterioration, weathering, durability and degradation of materials and products

ES — emergency standards.

A sequential number follows the letter. If the standard is written using metric units but has a companion standard in inch-pound units (or any other type of units), it is then followed by an M to identify the metric standard. Next, usually following a hyphen, is either the full year in which the standard was issued or else the last two numbers of that year (89 for 1989). Some organizations will change this number to indicate the year in which the standard was last revised; others place this information in parenthesis after the title of the standard. Should the year be followed by a lower case letter, it indicates that there was more than one revision of the standard during that year ("a" indicates the second revision, "b" indicates the third, etc.).

As you can see, using the information in the previous paragraph, the number itself provides an enormous amount of information. We know just by looking at it that the standard to which it refers is issued by the American Society for Testing and Materials, that it deals with an end-use material or product, and that it is written using metric units.

Indexes of Standards: There are many sources available to locate the necessary standards for any design project. While they do not hold in-depth information on individual standards, they do provide information relevant to many topics. Some of the information available in these references includes lists of standards-related organizations, cumulative subject indexes of standards (giving related organizations, the standards themselves, or standards-related periodicals), and numeric lists of standards. Also available are searchable indexes that can be found on many standards-issuing organizations' web pages. They allow users to search by keywords in the standard or by its number. While these pages do not allow access to the details contained in the standards, they do give the number and title of the standard. In some cases they contribute a brief description of the type of information included.

Standards covered here: There are an incredible number of standards-issuing organizations throughout the world. In this introduction we summarize several of those which issue standards pertinent to design in Mechanics of Materials. They are arranged in alphabetical order by the organization's name. Topics range from materials used in design to ergonomics.

Aerospace Industries Association of America, Inc. http://www.access.digex.net/~aia/ AIA represents America's leading manufacturers of commercial, military, and business aircraft, helicopters, aircraft engines, missiles, spacecraft, and related components and equipment. AIA provides many of the country's national aerospace standards, which cover such topics as screws, bolts, nuts and washers; wires; tube assembly; twin seaplane floats; shackle components for ground support equipment; angles; tees and brackets.

The Aluminum Association, Inc. http://www.aluminum.org/ AAI is the trade association for U.S. producers of aluminum and semi-fabricated aluminum products as well as aluminum recyclers. AAI provides leadership to the industry through its programs and services which aim to increase the use of aluminum, remove impediments to its advancement and assist in achieving the industry's environmental, societal, and economic objectives. AAI publishes standards for aluminum and its alloys for various shapes and types of manufacturing, as well as for aluminum design in general.

Topics Covered Include

Minimum and typical mechanical properties for regular and welded aluminum alloys

Allowable stresses for bridge type structures

Allowable stresses for building and similar type structures

Allowable uniform beam loads for given depth, weight, and span

Allowable stresses for mechanical connections

Dimensions for various types of bolts, nuts and washers

Comparative characteristics and applications of aluminum alloys.

Also given is information such as designation, sizes, weights, moment of inertia and radius of gyration for many different types of channels, flanges, I-beams, angles, tee and zee sections, round tubes, pipe, square tubes, and rectangular tubes.

There are references available, to be used with AAI's standards, which give in-depth explanations, both qualitative and quantitative, on how to analyze aluminum elements used in design.

The American Institute of Steel Construction http://www.aisc.org/ AISC represents and serves the structural steel industry in the U.S. Its purpose is to use research and development, education, technical assistance, standardization and quality control to expand the use of fabricated structural steel. The standards, it publishes deal with steel including:

Steel beams

Composite beams

Steel connections

Allowable stress design in steel construction

Standard practices for steel buildings and bridges

Allowable stress design specifications for structural joints

Specifications for structural steel buildings as pertaining to the allowable stress design of single angle members

Load and resistance factor design of simple shear connections and moment shears

Reactions for continuous highway bridges

The American Iron and Steel Institute http://www.steel.org/ AISI promotes the use of steel by providing standards for high-quality products, steel production, safety and environmental responsibility. AISI develops standards and publications regarding:

Critical issues of automotive steels - weight, cost, safety, recycling, and manufacturing

Shear resistance of walls with steel studs

Welded steel pipe and steel plate engineering data

Properties of bridge steels

Design and fabrication of cold-formed steel structures.

The American National Standards Institute http://web.ansi.org/default_js.htm. ANSI serves as the administrator and coordinator of the United States private sector voluntary standardization system and promotes and facilitates voluntary consensus standards and conformity assessment systems. ANSI itself does not develop American National Standards, it facilitates development through the establishment of consensus among qualified groups, who are accredited under one of its three methods; organization, committee, or by voting. ANSI is also the only U.S. representative to two major international standards organizations, the International Organization for Standardization, where it is one of only five permanent members on the governing ISO Council, and the International Electrotechnical Commission (IEC).

The American Society of Civil Engineers http://www.asce.org/ ASCE is America's oldest national engineering society. It seeks to advance professional knowledge and improve the practice of civil engineering by serving as the leading professional organization supporting both civil engineers and those in related fields. It also serves as the focal point for the development and transfer of research results and technical, managerial and policy related information and as a catalyst for effective and efficient service through cooperation with other engineering and related organizations. More than 6000 engineers serve on over 580 national committees that produce the annual convention, specialty conferences, publications, policies, and building codes and standards, among other services. Standards for structural design of composite slabs, minimum design loads for buildings and other structures, specifications for the design of cold-formed stainless steel structural members, and the design of latticed steel transmission structures are a few of those issued by ASCE.

The American Society of Mechanical Engineers http://www.asme.org/ ASME International has nearly 600 codes and standards in print for the design, manufacturing, and installation of mechanical devices. The development of such codes conforms to the procedures set by the American National Standards Institute. ASME standards deal with every possible element of mechanical engineering from boilers and pressure vessels to fluid flow and piping.

Among those which will most likely be of use to students in an introductory mechanics course, such as Strength of Materials, are standards dealing with:

Bolts, screws and nuts of both the square and hex variety

Lock and plain washers

Standard dimensions, limitations, and length tolerances

Pipes, fittings, flanges, gaskets, saw blades, socket wrenches, and monorails

Power transmission through various kinds of chains

Cranes

Pressure vessels and codes for their manufacture

The American Society for Testing and Materials http://www.astm.org/ ASTM is one of the largest voluntary standards developers in the world, providing a forum for producers, users, ultimate consumers, and those having a general interest to meet and write standards for materials, products, systems, and services. All technical research and testing is done voluntarily by technically qualified ASTM members throughout the world. ASTM develops six main types of full consensus standards:

Standard test methods—procedures for the identification, measurement, and evaluation of one or more qualities, characteristics or properties of a material, product, system or service

Standard specifications—precise statements of a set of requirements a material, product, system or service needs to satisfy and the procedures for determining whether the requirements are satisfied

Standard practices—procedures for performing one or more specific operation or function that does not give a test result

Standard terminology—documents of terms, definitions, descriptions of terms, explanations of symbols, abbreviations or acronyms

Standard guides—series of options or instructions

Standard classifications—systematic arrangements or divisions of materials, products, systems or services into groups based on similar characteristics.

ASTM's standards development activities cover many areas including metals, paints, plastics, textiles, petroleum, construction, energy, the environment, consumer products, medical services and devices, computerized systems and electronics. Thus the volumes it publishes hold an incredible amount of information, so an alphabetical subject index is also provided to make it easier for users to find those standards which they desire. Some specific examples of the topics included are: bicycle child carriers, high chairs, hook-on chairs, plastic chairs for outdoor use, various types of playground equipment (swings, slides, merry-go-rounds, etc), many different fasteners, steel piping, steel tubing and fitting, plastic pipe and building products and aluminum alloys of various shapes and manufacturing types.

Building Officials and Code Administrators International, Inc. http://www.bocai.org Boca codes include primarily those standards which deal with the construction of buildings and the components which go into their design, such as fire protection and the environmental systems (heating, ventilating, and air-conditioning).

There is some information, however, which may be useful to students who are in an introductory mechanics course. This information includes deflection standards for reinforced concrete, structural steel, masonry, roofs, walls, and floors as well as information on building materials such as concrete, steel, wood, glass and glazing, gypsum board and plaster, and plastic. There is also information regarding the quantity and size of fasteners connecting wood frame members together as well as different types of structural loads (snow, wind, and earthquake).

Ergonomics Standards http://www.ergoweb.com/ http://www.interface-analysis.com/ergoworld/ The purpose of ergonomics standards is to provide information that engineers and designers need to make equipment fit the human body. Ergonomics, or human factors engineering, takes into account properties of the mind and body as they are manifested in people's interaction with the environment. To help designers accomplish this, many sources provide information required to make a project ergonomic.

There is information available concerning:

Dimensions of the human body in several different work positions

Information on the skeletal, muscular, respiratory and circulatory systems

Human factors standards used to design specific items, such as hand tools and furniture.

The International Standards Organization http://www.iso.ch/welcome.html ISO, properly known as The International Organization for Standardization, is a worldwide federation of national standards bodies from about 100 countries, with one representative from each country.

ISO works to promote the development of standardization in the world with the goal of improving the international exchange of goods and services and to develop cooperation in the areas of intellectual, scientific, technological, and economic activity.

The member bodies of ISO have four principle tasks:

Informing potentially interested parties in their country of relevant international standardization opportunities and initiatives

Organizing so that a concerted view of the country's interests is presented during international negotiations leading to standards agreements

Ensuring that a secretariat is provided for those ISO technical committees and subcommittees in which the country has an interest

Providing their country's share of financial support for the central operations of ISO ISO standards are numerous and cover an incredibly broad range of topics, from the optimum width of an ATM card to the load ratings, dimensions, types, parts and subunits of rolling bearings.

Perhaps their most renowned standard is the ISO 9000 series, which deals with quality control.

Other examples of topics ISO covers are acoustics, vibration, shock, hand tools (such as files, rasps, hand reamers, drills, wrenches and sockets, abrasive sheets, screwdrivers, and pliers), machine tools (T-slots and corresponding bolts, modular units for machine tool construction, woodworking machines, and lubricating systems), mechanical transmission and applied metrology.

The Society of Automotive Engineers http://www.sae.org/ SAE is made up of engineers, business executives, educators and students from all over the world who provide networking opportunities, share information and exchange ideas for advancing the engineering of mobility systems. SAE technical committees write more aerospace and automotive engineering standards than any other standards-writing organization in the world.

These standards are grouped into three main categories: ground vehicle standards (J-Reports), aerospace standards and aerospace material specifications (AMS). More specifically, topics covered include:

Molded rigid plastic parts.

Bolts, nuts, washers and screws.

Human physical dimensions.

Mechanical and physical properties of aluminum alloys, cobalt alloys, composite materials, copper alloys, copper tubing, copper-nickel tubing, felt, iron alloys, lead alloys, magnesium alloys, nickel alloys, plastics, rubber, steel (bars, pipe, shot, springs, wire, tubing), tin alloys, titanium alloys and zinc alloys

Dimensions of curbstone clearance and human tolerance to impact conditions as related to motor vehicle design.

Other Organizations

The American Association of State Highway and Transportation Officials (AASHTO) focuses on those standards dealing with highway and bridge construction. http://www.aashto.org/

The American Concrete Institute (ACI) develops standards dealing primarily with concrete. They provide information such as the erosion of concrete under various circumstances, corrosion of metals in concrete, cracking of concrete members in direct tension, shear strength of reinforced concrete members, allowable deflections, deflections of prestressed concrete members and control of deflections in concrete. http://www.aci-int.org/

The American Forest and Paper Association (AFPA), formerly the National Forest Products Association (NFoPA), is the national trade association of the forest, paper, and wood products industry. Its standards deal with wood structural design and include general requirements, design provisions and formulas, and data on sawn lumber, structural glued laminated timber, round timber piles, and connection and design values for sawn lumber and glued laminated timber. http://www.afandpa.org/

The Association of Iron and Steel Engineers (AISE) is the largest member-based technical organization in the steel industry. It writes standards covering such topics as alloy steel chain and alloy steel chain slings for overhead lifting, brake standards for mill motors, specifications for electric traveling cranes for steel mill service, specifications for ladle hooks, and a guide for the design and construction of mill buildings. http://www.aise.org/

What are Codes?

A code is a set of rules and specifications for the correct methods and materials used in a certain product, building or process. Codes can be approved by local, state or federal governments and can carry the force of law. The main purpose of codes is to protect the public by setting up the minimum acceptable level of safety for buildings, products and processes.

Product Safety and Liabilities

Basic concepts of "safety" are changing around the world. Product liability regimes and especially product safety regimes have traditionally focused on risk of physical injury and, to a lesser extent, risks of property damage. However, the concept of safety is expanding and companies are having to adapt their own compliance and risk management systems accordingly.

Increasingly, the concept of a safe product in various contexts takes into account concepts of environmental protection, data privacy and even (albeit tangentially) energy efficiency. It also, more and more, needs to take into account both "end-of-life" safety and concepts of "foreseeable misuse" as well as intended use. These are all challenging issues for product manufacturers.

Whilst many countries around the world have had robust laws and regulations in place to help ensure product safety for some time, it has been a common feature in most parts of the world that enforcement has been sporadic or ineffective. This has been well recognised in many countries, where concerns are raised about the lack of funding allocated by governments to product safety enforcement.

This issue has been raised more prominently in the eyes of the public in recent years through some high-profile international regulatory scandals that have caused some to lose trust in regulators as well as in companies.

We are now in a world where we are seeing greater emphasis amongst policy-makers and regulators on finding more effective ways to enforce laws and regulations. In some cases, this involves empowering consumers and other claimants to bring claims more effectively against companies; in others, ensuring enforcement agencies have better resources, not only in terms of funding but also in terms of powers and information. This is an important factor that is set to significantly alter the risks being managed by international product manufacturers and suppliers moving forward.

Resources for Managing Changing Risks

The increasing complexity and risks of the product law world are a significant source of new challenges for product manufacturers and suppliers who seek to take advantage of global markets. The costs of failing to understand, anticipate and manage the risks can be high, as many high-profile brands have discovered in recent years.

On the other hand, these developments also create opportunities.

The development of rules and regulations, together with the emergence of more active enforcement agencies, can help to ensure a level playing field and stable markets for companies that have an interest in ensuring they comply with the rules. Companies with valuable brand names and reputations to protect can be especially exposed when marketing their products in markets that have few controls. Proportionate laws, fairly and effectively enforced, can help companies to fully realise the benefits of their investments and manage their risks.

The key is for companies to find those practical ways to keep abreast of the changes, understand their implications and develop systems that are fit for the future.

Product safety problem has appeared many years in the developed countries, but it was just begun to take seriously in recent years in our country. The product liability systems which established with the development of technology and economy reflect the concept of product safety of human being. Foreign academics have analyzed the problem of product liability system, the effect of product liability to product and process design, etc.

There was little study has been done in our country in the concept of product liability, not to mention its effect to inherent safety of industry. Product liability acts have been issued in many developed countries, and some laws have been established though there was not special act in product liability in our country until now. But accidents caused by product defects increased in recent years. The product liability system did not play a role efficiently. It is significance to analysis the product liability concept and the problems exist in the product liability laws to search for specific countermeasures to prevention accidents.

Development of Product Liability Concept

Product liability is the legal responsibility of the product maker or sellers and other relevant main body that lead other people's physical or property damage (not including the damage of defect product itself) because of the defect product.

Product liability appeared with the development of modern industry, and its doctrine of liability fixation changed with the economy and technical development constantly. It influenced the manner of enterprise's product liability prevention, and also affected inherent safety of industrial production.

PRODUCT SAFETY, LIABILITY, AND ENGINEERING ETHICS

During the same period as the Carson and Nader books, professional engineering societies began to take more seriously the role of engineers and the engineering profession as stewards of product safety. All contemporary codes of engineering ethics state that engineers have a responsibility to protect the public safety, health, and welfare, and most codes state that this duty should be held paramount.

The notion that safety is of primary importance in engineering is also fundamental to nearly all academic treatments of engineering ethics (Herkert 2000). A key concept is the notion of professional responsibility, which many ethicists characterize as a type of moral responsibility arising from special knowledge possessed by an individual. Philosopher Mike Martin and engineer Roland Schinzinger argue that professional responsibility in engineering involves "the creation of useful and safe technological products while respecting the autonomy of clients and the public, especially in matters of risk-taking".

Yet while product safety is central to discussions of engineering ethics, the closely related legal concept of product liability is often ignored, or even attacked by engineering professionals and others. "Developing from the Industrial Revolution, U.S. product liability law is derived from case law and restatements of law anchored in contract and tort. It is based on the belief that consumers need protection from business and that business should bear the costs of harms inflicted on consumers".

Over time, the legal standard regarding product liability has evolved from the doctrine of let the buyer beware, to a legal theory requiring a determination of negligence on the part of the manufacturer, to the modern legal standard of strict liability (liability imposed without fault). Product liability claims can be based on manufacturing defects, design defects, and information defects (lack of appropriate warnings).

Judgments in product liability cases can include both compensatory (reimbursement for costs) and punitive damages; large judgments have often been the focus of attention in the controversy over product liability, especially in cases when the judgment may seem out of proportion to the harm. In one notorious case, a jury awarded a woman nearly $3 million for burns she received when she spilled coffee purchased at a McDonald's drive-up window.

Critics of current product liability law, including many professional engineering societies, call for rollbacks often approaching the old let the buyer beware policies. For example, in 1996 Congress passed legislation that would have severely limited the effect of product liability litigation by placing a cap on punitive damages and enacting stricter requirements for holding manufacturers liable. President Bill Clinton vetoed the bill; however, the debate over product liability reform continued. The proponents of product liability reform argue that the current system unjustly rewards plaintiffs and stifles technological innovation, resulting in a lack of competitiveness on the part of U.S. manufacturers and decreased product safety. Supporters of the current system counter that it generally works as intended in discouraging the manufacture of defective products and compensating people injured by such defects. To some the debate over product liability reform is a classic business/consumer conflict. A New York Times editorial (1996), for example, described proposed legislation as "The Anti-Consumer Act of 1996." Despite the arguments of both sides, the evidence is mixed concerning whether product liability rewards result in improvements in product safety.

Engineers and engineering societies have tended to side with the proponents of product liability reform. A vice president of engineering of a major U.S. automobile company, for example, has argued that product liability restricts engineering practice by inhibiting innovation, discouraging critical evaluation of safety features, and preventing implementation of new or improved designs. The 1998 position statement on product liability of IEEE-USA, a unit of the Institute of Electrical and Electronics Engineers (IEEE) concerned with professional issues in the United States, calls for stringent limits on product liability including holding the manufacturer blameless when existing standards are met, adequate warnings are provided, or the product is misused or altered by the user. Other engineering societies, such as ASME International (formerly the American Society of Mechanical Engineers) have also actively supported product liability reform.

Given the primary responsibility of engineers for public safety, health, and welfare stated in the codes of ethics, it is surprising that the product liability issue has not drawn more attention from the perspective of engineering ethics. There is little, if any, evidence, however, to suggest that engineering societies promoting changes in the product liability system have considered the effect that decreasing the impact of product liability would have from the point of view of engineering ethics.

On the whole, the engineering community has paid little attention to the ethical implications of product liability. For example, a major study of product liability and innovation by the National Academy of Engineering which considered such issues as corporate practice, insurance, regulation, and the role of scientific and technical information in the courtroom, touched only briefly on ethics. Even the ethics literature is equivocal on the issue of product liability. For example, one well-known essay on engineering responsibility in the Ford Pinto case advocated stronger regulation and fines and imprisonment for corporate officials to achieve desired levels of safety, giving only passing notice to the role of product liability litigation.

Design Factors and Factor of Safety

The difference between the safety factor and design factor (design safety factor) is as follows: The safety factor, or yield stress, is how much the designed part actually will be able to withstand (first "use" from above). The design factor, or working stress, is what the item is required to be able to withstand (second "use"). The design factor is defined for an application (generally provided in advance and often set by regulatory building codes or policy) and is not an actual calculation, the safety factor is a ratio of maximum strength to intended load for the actual item that was designed.

Design load being the maximum load the part should ever see in service.

By this definition, a structure with an FOS of exactly 1 will support only the design load and no more. Any additional load will cause the structure to fail. A structure with an FOS of 2 will fail at twice the design load.

The following considerations can help design engineers account for real-world conditions, and create a better product.

Intensity of stress concentrations; which components will be subjected to more intense stress, more often?

Is thermal cycling or extreme temperature exposure an issue? How does this impact function, or the materials to be used?

Is scheduled maintenance likely to occur? Will it happen regularly, and be of similar quality each time?

Does the combination of materials used weaken or strengthen the overall design?

Is it likely that wear and tear will be accelerated through consumer use (say, regularly exceeding rated capacity, etc)? Are there controls in place to help prevent this?

Is the structure or component subject to deterioration through corrosion?

Margin of Safety

Many government agencies and industries (such as aerospace) require the use of a margin of safety (MoS or M.S.) to describe the ratio of the strength of the structure to the requirements. There are two separate definitions for the margin of safety so care is needed to determine which is being used for a given application.

One usage of M.S. is as a measure of capability like FoS. The other usage of M.S. is as a measure of satisfying design requirements (requirement verification). Margin of safety can be conceptualized (along with the reserve factor explained below) to represent how much of the structure's total capability is held "in reserve" during loading. M.S. as a measure of structural capability: This definition of margin of safety commonly seen in textbooks describes what additional load beyond the design load a part can withstand before failing. In effect, this is a measure of excess capability. If the margin is 0, the part will not take any additional load before it fails, if it is negative the part will fail before reaching its design load in service. If the margin is 1, it can withstand one additional load of equal force to the maximum load it was designed to support (i.e. twice the design load).

M.S. as a measure of requirement verification: Many agencies and organizations such as NASA and AIAA define the margin of safety including the design factor, in other words, the margin of safety is calculated after applying the design factor. In the case of a margin of 0, the part is at exactly the required strength (the safety factor would equal the design factor). If there is a part with a required design factor of 3 and a margin of 1, the part would have a safety factor of 6 (capable of supporting two loads equal to its design factor of 3, supporting six times the design load before failure). A margin of 0 would mean the part would pass with a safety factor of 3. If the margin is less than 0 in this definition, although the part will not necessarily fail, the design requirement has not been met. A convenience of this usage is that for all applications, a margin of 0 or higher is passing, one does not need to know application details or compare against requirements, just glancing at the margin calculation tells whether the design passes or not. This is helpful for oversight and reviewing on projects with various integrated components, as different components may have various design factors involved and the margin calculation helps prevent confusion.

Design safety factor = [provided as requirement]

For a successful design, the realized safety factor must always equal or exceed the design safety factor so that the margin of safety is greater than or equal to zero. The margin of safety is sometimes, but infrequently, used as a percentage, i.e., a 0.50 M.S is equivalent to a 50% M.S. When a design satisfies this test it is said to have a "positive margin", and, conversely, a "negative margin" when it does not.

Dimensions and Tolerances

Proper dimensioning is the key to obtaining what is wanted at optimum cost. Proper dimensioning will assure a precise product, speed production and permit accurate measurement. It is probably the most important phase in preparing a drawing. Dimensions serve two important functions. First, they limit the size of the part shown on the drawing. Second, they define the limits within which the parts will be acceptable. Precise dimensions correctly applied aid the user as well as the producer.

Tolerances should be noted for each dimension, and a tolerance block should note those that apply to all other dimensions. When noting tolerances or dimensions, it is preferred that fractions not be used and that a decimal system be used to show all tolerances. Specify tolerances only as close as necessary for part functions — but clearly state a tolerance for all lineal and radial dimensions and angles.

Surface texture designations should conform to ANSI/ASME Y14.36M, Surface Texture Symbols. Unless a maximum and minimum value is shown, the value shown will be the average surface texture value for that surface as defined in this standard. A dimension shown only as a maximum implies no minimum. A corner, for example, shown as break ".005 maximum" could be supplied razor sharp and be to specification. If this is not what is wanted, a minimum must be shown.

Always dimension with this question in mind — "How can this be measured?" Try to take advantage of readily available stock size materials, tools and gages. Wherever possible, use a common reference datum point or plane for all dimensions. Be sure all distances are clearly dimensioned. Ambiguity leads to misinterpretation and trouble.

Show all dimensional relationships for concentricity, squareness or control of another dimension. Use standard symbols. Dimension in a flat plane view and use as many views as necessary to show each detail. Avoid perspective drawings. Threads are best dimensioned in the standard accepted manner; e.g. 1/4-20 UNC-2A. If special or restricted major, minor or pitch diameters are required, the maximum and minimum limits should be dimensioned in decimals to three or more places. A chamfer at the end of a thread should be dimensioned at least.015" below the thread root diameter.

GD&T

GD&T, short for Geometric Dimensioning and Tolerancing, is a system for defining and communicating design intent and engineering tolerances that helps engineers and manufacturers optimally control variations in manufacturing processes.

Limitations of Tolerancing Before GD&T

Before GD&T, manufacturing features were specified by X-Y areas. For example, when drilling a mounting hole, the hole had to be within a specified X-Y area. An accurate tolerancing specification, however, would define the position of the hole in relation to the intended position, the accepted area being a circle. X-Y tolerancing leaves a zone in which inspection would have produced a false negative because while the hole is not within the X-Y square, it would fall within the circumscribed circle.

Stanley Parker, an engineer who was developing naval weapons during World War II, noticed this failure in 1940. Driven by the need for cost-effective manufacturing and meeting deadlines, he worked out a new system through several publications. Once proven as a better operational method, the new system became a military standard in the 1950s.

Currently, the GD&T standard is defined by the American Society of Mechanical Engineers (ASME Y14.5-2018) for the USA and ISO 1101-2017 for the rest of the world. It concerns mostly the overall geometry of the product, while other standards describe specific features such as surface roughness, texture, and screw threads.

Why Implement GD&T Processes?

With functional assemblies, multi-part products, or parts with complex functionality, it is crucial that all components work well together. All relevant fits and features need to be specified in a way that impacts the manufacturing process and its related investments the least, while still guaranteeing functionality. Tightening tolerances by a factor two can raise the costs twofold or even more, due to higher reject rates and tooling changes. GD&T is the system that allows developers and inspectors to optimize functionality without increasing cost.

The most important benefit of GD&T is that the system describes the design intent rather than the resulting geometry itself. Like a vector or formula, it is not the actual object but a representation of it.

For example, a feature standing at 90 degrees to a base surface can be toleranced on its perpendicularity to that surface. This will define two planes spaced apart, that the center plane of the feature must fall within. Or, when drilling a hole, it makes the most sense to tolerance it in terms of alignment to other features. Describing product geometry related to its intended functionality and manufacturing approach is ultimately simpler than having to describe everything in linear dimensions. It also provides a communication tool with manufacturing vendors, customers, as well as quality inspectors.

When performed well, GD&T even allows statistical process control (SPC), reducing product reject rates, assembly failures, and the effort needed for quality control, saving organizations substantial resources. As a result, multiple departments are able to work more in parallel because they have a shared vision and language for what they want to achieve.

How GD&T Works?

Engineering drawings need to show the dimensions for all features of a part. Next to the dimensions, a tolerance value needs to be specified with the minimum and maximum acceptable limit. The tolerance is the difference between the minimum and maximum limit. For example, if we have a table that we would accept with a height between 750 mm and 780 mm, the tolerance would be 30 mm.

However, the tolerance for the table implies that we would accept a table that is 750 mm high on one side and 780 mm on the other, or has a waved surface with 30 mm variation. So to appropriately tolerance the product, we need a symbol communicating the design intent of a flat top surface. Therefore we have to include an additional flatness tolerance in addition to the overall height tolerance.

Similarly, a cylinder with a toleranced diameter will not necessarily fit into its hole if the cylinder gets slightly bent during the manufacturing process. Therefore it also needs a straightness control, which would be difficult to communicate with traditional plus-minus tolerancing. Or a tube that has to seamlessly match a complex surface that it's welded to requires a surface profile control.

The art of tolerancing means to specify just the right variations for all specific design features in order to maximize product approval rate within the limits of the manufacturing processes and depending on the part's visual and functional purpose.

In the metric system, there are International Tolerance (IT) grades that can also be used to specify tolerances by means of symbols. The symbol 40H11, for example, means a 40 mm diameter hole with a loose running fit. The manufacturer then only needs to look up the basis table for hole features to derive the exact tolerance value.

Besides individual tolerances, engineers must take into account system-level effects. For example, when a part comes out with all dimensions at their maximum allowed value, does it still meet overall requirements such as product weight and wall thicknesses? This is called the Maximum Material Condition (MMC), while its counterpart is the Least Material Condition (LMC). Tolerances also stack up. If we create a chain link where each hole has a 0.1 mm plus tolerance and each shaft a 0.1 mm negative tolerance, that means we will still accept a 20 mm length difference at

100 links. When installing repeated elements such as a perforated hole pattern, first position the pattern and then specify interrelated distances rather than referencing elements to a fixed edge or plane of the part.

The standards do not only pertain to designers and engineers but also to quality inspectors by informing them how to measure the dimensions and tolerances. Using specific tools such as digital micrometers and calipers, height gauges, surface plates, dial indicators, and a coordinate measuring machine (CMM) are important to tolerancing practice.

When measuring and defining a part, the geometry exists in a conceptual space called the Datum Reference Frame (DRF). This is comparable to the coordinate system at the origin of a space in 3D modeling programs. A datum is a point, line or plane that exists in the DRF and is used as a starting place for measuring. Make sure to define the datum features relevant to the functionality of your part. Unless you are mating features of one part to those of others in an assembly, you can often use a single datum. Always make sure that the primary datum has a reliable location to derive other measurements from, for example, where the final part will have little unpredictable variation.

GD&T Tolerancing Guidelines

An engineering drawing has to accurately convey the product without adding unnecessary complexity or restrictions. The following guidelines are helpful to consider:

Clarity of a drawing is the most important, even more so than its accuracy and completeness. To improve clarity, draw dimensions and tolerances outside of the part's boundaries and applied to visible lines in true profiles, employ a unidirectional reading direction, convey the function of the part, group and/or stagger dimensions, and make use of white space.

Always design for the loosest feasible tolerance to keep costs down.

Use a general tolerance defined at the bottom of the drawing for all dimensions of the part. Specific tighter or looser tolerances indicated in the drawing will then supersede the general tolerance.

Tolerance functional features and their interrelations first, then move on to the rest of the part.

Whenever possible, leave GD&T work to the manufacturing experts and do not describe manufacturing processes in the engineering drawing.

Do not specify a 90-degree angle since it is assumed.

Dimensions and tolerances are valid at 20 °C / 101.3 kPa unless stated otherwise.

GD&T is Feature-based, with Each Feature Specified by Different Controls.

These tolerancing symbols fall into five groups:

Form controls specify the shape of features, including:

Straightness is divided into line element straightness and axis straightness.

Flatness means straightness in multiple dimensions, measured between the highest and lowest points on a surface.

Circularity or roundness can be described as straightness bent into a circle.

Cylindricity is basically flatness bent into a barrel. It includes straightness, roundness, and taper, which makes it expensive to inspect.

Profile controls describe the three-dimensional tolerance zone around a surface:

Line Profile compares a two-dimensional cross-section to an ideal shape. The tolerance zone is defined by two offset curves unless otherwise specified.

Surface Profile creates through two offset surfaces between which the feature surface must fall. This is a complex control typically measured with a CMM.

Orientation controls concern dimensions that vary at angles, including:

Angularity is flatness at an angle to a datum and is also determined through two reference planes spaced the tolerance value apart.

Perpendicularity means flatness at 90 degrees to a datum. It specifies two perfect planes the feature plane must lie in between.

Parallelism means straightness at a distance. Parallelism for axes can be defined by defining a cylindrical tolerance zone by placing a diameter symbol in front of the tolerance value.

Location controls define feature locations using linear dimensions:

Position is the location of features relative to one another or to datums and is the most used control.

Concentricity compares the location of a feature axis to the datum axis.

Symmetry ensures that non-cylindrical parts are similar across a datum plane. This is a complex control typically measured with a CMM.

Runout controls define the amount by which a particular feature can vary with respect to the datums:

Circular Runout is used when there is a need to account for many different errors, such as ball-bearing mounted parts. During inspection, the part is rotated on a spindle to measure the variation or 'wobble' around the rotational axis.

Total Runout is measured on multiple points of a surface, not just describing the runout of a circular feature but of an entire surface. This controls straightness, profile, angularity, and other variations.

Power Transmission Case Study Specifications

Many industrial applications require machinery to be powered by engines or electric motors. The power source usually runs most efficiently at a narrow range of rotational speed. When the application requires power to be delivered at a slower speed than supplied by the motor, a speed reducer is introduced.

The speed reducer should transmit the power from the motor to the application with as little energy loss as practical, while reducing the speed and consequently increasing the torque. For example, assume that a company wishes to provide off-the-shelf speed reducers in various capacities and speed ratios to sell to a wide variety of target applications. The marketing team has determined a need for one of these speed reducers to satisfy the following customer requirements.

Design Requirements

Power to be delivered: 20 hp.

Input speed: 1750 rev/min.

Output speed: 85 rev/min.

Targeted for uniformly loaded applications, such as conveyor belts, blowers, and generators.

Output shaft and input shaft in-line.

Base mounted with 4 bolts.

Continuous operation.

6-year life, with 8 hours/day, 5 days/wk.

Low maintenance.

Competitive cost.

Nominal operating conditions of industrialized locations.

Input and output shafts standard size for typical couplings.

In reality, the company would likely design for a whole range of speed ratios for each power capacity, obtainable by interchanging gear sizes within the same overall design. For simplicity, in this case study only one speed ratio will be considered. Notice that the list of customer requirements includes some numerical specifics, but also includes some generalized requirements, e.g., low maintenance and competitive cost. These general requirements give some guidance on what needs to be considered in the design process, but are difficult to achieve with any certainty.

In order to pin down these nebulous requirements, it is best to further develop the customer requirements into a set of product specifications that are measurable. This task is usually achieved through the work of a team including engineering, marketing, management, and customers.

Various tools may be used to prioritize the requirements, determine suitable metrics to be achieved, and to establish target values for each metric. The goal of this process is to obtain a product specification that identifies precisely what the product must satisfy. The following product specifications provide an appropriate framework for this design task.

Design Specifications

Power to be delivered: 20 hp

Power efficiency: >95%

Steady state input speed: 1750 rev/min

Maximum input speed: 2400 rev/min

Steady-state output speed: 82–88 rev/min

Usually low shock levels, occasional moderate shock

Input and output shaft diameter tolerance: ±0.001 in

Output shaft and input shaft in-line: concentricity ±0.005 in, alignment ±0.001 rad

Maximum allowable loads on input shaft: axial, 50 lbf; transverse, 100 lbf

Maximum allowable loads on output shaft: axial, 50 lbf; transverse, 500 lbf

Base mounted with 4 bolts

Mounting orientation only with base on bottom 100% duty cycle

Maintenance schedule: lubrication check every 2000 hours; change of lubrication every 8000 hours of operation; gears and bearing life >12,000 hours; infinite shaft life; gears, bearings, and shafts replaceable

Access to check, drain, and refill lubrication without disassembly or opening of gasketed joints.

Manufacturing cost per unit: <$300

Production: 10,000 units per year

Operating temperature range: “10æ% to 120æ%F

Sealed against water and dust from typical weather

Noise: <85 dB from 1 meter

BIBLIOGRAPHY

- Ahmad Y Hassan, Donald Routledge Hill (1986). Islamic Technology: An illustrated history, p. 54. Cambridge University Press. ISBN 0-521-42239-6.
- Ahmad Y. Hassan (1976), Taqi al-Din and Arabic Mechanical Engineering, p. 34-35, Institute for the History of Arabic Science, University of Aleppo
- Arnold, Dieter (1991). Building in Egypt: Pharaonic Stone Masonry. Oxford University Press. p. 71. ISBN 9780195113747.
- Burstall, Aubrey F. (1965). A History of Mechanical Engineering. The MIT Press. ISBN 978-0-262-52001-0.
- Collins, Robert O.; Burns, James M. (8 February 2007). A History of Sub-Saharan Africa. Cambridge University Press. ISBN 9780521867467 – via Google Books.
- Depuydt, Leo (1 January 1998). Gnomons at Meroë and Early Trigonometry. The Journal of Egyptian Archaeology. 84: 171–180. doi:10.2307/3822211. JSTOR 3822211.
- Donald Routledge Hill, Mechanical Engineering in the Medieval Near East, Scientific American, May 1991, pp. 64–9 (cf. Donald Routledge Hill, Mechanical Engineering)
- Edwards, David N. (29 July 2004). The Nubian Past: An Archaeology of the Sudan. Taylor & Francis. ISBN 9780203482766 – via Google Books.
- Eldridge, Frank (1980). Wind Machines (2nd ed.). New York: Litton Educational Publishing, Inc. p. 15. ISBN 0-442-26134-9.
- Episode 11: Ancient Robots, Ancient Discoveries, History Channel, retrieved 2008-09-06

- G. Mokhtar (1981-01-01). Ancient civilizations of Africa. Unesco. International Scientific Committee for the Drafting of a General History of Africa. p. 309. ISBN 9780435948054. Retrieved 2012-06-19 – via Books.google.com.
- Gimpel, Jean (1976). The Medieval Machine : The Industrial Revolution of the Middle Ages. pp. 1–24, 66–67. ISBN 9780030146367.
- Heron of Alexandria. Encyclopædia Britannica 2010 - Encyclopædia Britannica Online. Accessed: 9 May 2010.
- Howard R. Turner (1997), Science in Medieval Islam: An Illustrated Introduction, p. 184, University of Texas Press, ISBN 0-292-78149-0
- Humphris J, Charlton MF, Keen J, Sauder L, Alshishani F (June 2018). Iron Smelting in Sudan: Experimental Archaeology at The Royal City of Meroe. Journal of Field Archaeology. 43 (5): 399–416. doi:10.1080/00934690.2018.1479085.
- Humphris, Jane; Charlton, Michael F.; Keen, Jake; Sauder, Lee; Alshishani, Fareed (2018). Iron Smelting in Sudan: Experimental Archaeology at The Royal City of Meroe. Journal of Field Archaeology. 43 (5): 399. doi:10.1080/00934690.2018.1479085. ISSN 0093-4690.
- Kapur, Ajay; Carnegie, Dale; Murphy, Jim; Long, Jason (2017). Loudspeakers Optional: A history of non-loudspeaker-based electroacoustic music. Organised Sound. Cambridge University Press. 22 (2): 195–205. doi:10.1017/S1355771817000103. ISSN 1355-7718.
- Koetsier, Teun (2001), On the prehistory of programmable machines: musical automata, looms, calculators, Mechanism and Machine Theory, Elsevier, 36 (5): 589–603, doi:10.1016/S0094-114X(01)00005-2.
- Koldewey, Robert (1914). The excavations at Babylon. London: Macmillan and Co. p. 91. ISBN 9781298040022.
- Lakwete, Angela (2003). Inventing the Cotton Gin: Machine and Myth in Antebellum America. Baltimore: The Johns Hopkins University Press. pp. 1–6. ISBN 9780801873942.
- Lavan, Luke; Zanini, Enrico; Sarantis, Alexander (2007). Technology in Trainsition A.D. 300-650. Boston. pp. 373–374. ISBN 9789004165496.
- Lucas, Adam (2006), Wind, Water, Work: Ancient and Medieval Milling Technology, Brill Publishers, p. 65, ISBN 90-04-14649-0
- Marks' Standard Handbook for Mechanical Engineers (11 ed.). McGraw-Hill. 2007. ISBN 978-0-07-142867-5.

- Moorey, Peter Roger Stuart (1999). Ancient Mesopotamian Materials and Industries: The Archaeological Evidence. Eisenbrauns. p. 4. ISBN 9781575060422.
- Needham, Joseph (1986). Science and Civilization in China: Volume 4. Taipei: Caves Books, Ltd.
- Oberg, Erik; Franklin D. Jones; Holbrook L. Horton; Henry H. Ryffel; Christopher McCauley (2016). Machinery's Handbook (30th ed.). New York: Industrial Press Inc. ISBN 978-0-8311-3091-6.
- Pacey, Arnold (1991) [1990]. Technology in World Civilization: A Thousand-Year History (First MIT Press paperback ed.). Cambridge MA: The MIT Press. pp. 23–24.
- Professor Noel Sharkey, A 13th Century Programmable Robot (Archive), University of Sheffield.
- Selin, Helaine (2013). Encyclopaedia of the History of Science, Technology, and Medicine in Non-Westen Cultures. Springer Science & Business Media. p. 282. ISBN 9789401714167.
- Shepherd, William (2011). Electricity Generation Using Wind Power (1 ed.). Singapore: World Scientific Publishing Co. Pte. Ltd. p. 4. ISBN 978-981-4304-13-9.
- Slayman, Andrew (27 May 1998). Neolithic Skywatchers. Archaeology Magazine Archive. Archived from the original on 5 June 2011. Retrieved 17 April 2011.
- Taqi al-Din and the First Steam Turbine, 1551 A.D. Archived 2008-02-18 at the Wayback Machine, web page, accessed on line 23 October 2009; this web page refers to Ahmad Y Hassan (1976), Taqi al-Din and Arabic Mechanical Engineering, pp. 34-5, Institute for the History of Arabic Science, University of Aleppo.
- Wood, Michael (2000). Ancient Machines: From Grunts to Graffiti. Minneapolis, MN: Runestone Press. pp. 35, 36. ISBN 0-8225-2996-3.
- Woods, Michael; Mary B. Woods (2000). Ancient Machines: From Wedges to Waterwheels. USA: Twenty-First Century Books. p. 58. ISBN 0-8225-2994-7.
- Žmolek, Michael Andrew (2013). Rethinking the Industrial Revolution: Five Centuries of Transition from Agrarian to Industrial Capitalism in England. BRILL. p. 328. ISBN 9789004251793. The spinning jenny was basically an adaptation of its precursor the spinning wheel

INDEX